D0141765

Pollution

TREATING ENVIRONMENTAL TOXINS

Pollution

TREATING ENVIRONMENTAL TOXINS

Anne Maczulak, Ph.D.

Facts On File
An imprint of Infobase Publishing

POLLUTION: Treating Environmental Toxins

Facts On File, Inc.
An imprint of Infobase Publishing
132 West 31st Street
New York NY 10001

Library of Congress Cataloging-in-Publication Data
Maczulak, Anne E. (Anne Elizabeth), 1954–
Pollution: treating environmental toxins / Anne Maczulak.
p. cm.
Includes bibliographical references and index.
ISBN 978-0-8160-7202-6
1. Pollution. 2. Toxins. 3. Hazardous waste site remediation. I. Title.
TD174.M327 2010
363.738'45—dc22 2008055558

Facts On File books are available at special discounts when purchased in bulk quantities for businesses, associations, institutions, or sales promotions. Please call our Special Sales Department in New York at (212) 967-8800 or (800) 322-8755.

You can find Facts On File on the World Wide Web at http://www.factsonfile.com

Text design by James Scotto-Lavino
Illustrations by Bobbi McCutcheon
Photo research by Elizabeth H. Oakes

Printed in the United States of America

Bang Hermitage 10 9 8 7 6 5 4 3 2 1

This book is printed on acid-free paper.

Contents

Preface

The first Earth Day took place on April 22, 1970, and occurred mainly because a handful of farsighted people understood the damage being inflicted daily on the environment. They understood also that natural resources do not last forever. An increasing rate of environmental disasters, hazardous waste spills, and wholesale destruction of forests, clean water, and other resources convinced Earth Day's founders that saving the environment would require a determined effort from scientists and nonscientists alike. Environmental science thus traces its birth to the early 1970s.

Environmental scientists at first had a hard time convincing the world of oncoming calamity. Small daily changes to the environment are more difficult to see than single explosive events. As it happened the environment was being assaulted by both small damages and huge disasters. The public and its leaders could not ignore festering waste dumps, illnesses caused by *pollution,* or stretches of land no longer able to sustain life. Environmental laws began to take shape in the decade following the first Earth Day. With them, environmental science grew from a curiosity to a specialty taught in hundreds of universities.

The condition of the environment is constantly changing, but almost all scientists now agree it is not changing for the good. They agree on one other thing as well: Human activities are the major reason for the incredible harm dealt to the environment in the last 100 years. Some of these changes cannot be reversed. Environmental scientists therefore split their energies in addressing three aspects of ecology: cleaning up the damage already done to the Earth, changing current uses of natural resources, and developing new technologies to conserve Earth's remaining natural resources. These objectives are part of the green movement. When new technologies are invented to fulfill the objectives, they can collectively be called green technology. Green Technology is a multivolume set that explores new methods for repairing and restoring the environment. The set covers a broad range of subjects as indicated by the following titles of each book:

- *Cleaning Up the Environment*
- *Waste Treatment*
- *Biodiversity*
- *Conservation*
- *Pollution*
- *Sustainability*
- *Environmental Engineering*
- *Renewable Energy*

Each volume gives brief historical background on the subject and current technologies. New technologies in environmental science are the focus of the remainder of each volume. Some green technologies are more theoretical than real, and their use is far in the future. Other green technologies have moved into the mainstream of life in this country. Recycling, alternative energies, energy-efficient buildings, and biotechnology are examples of green technologies in use today.

This set of books does not ignore the importance of local efforts by ordinary citizens to preserve the environment. It explains also the role played by large international organizations in getting different countries and cultures to find common ground for using natural resources. Green Technology is therefore part science and part social study. As a biologist, I am encouraged by the innovative science that is directed toward rescuing the environment from further damage. One goal of this set is to explain the scientific opportunities available for students in environmental studies. I am also encouraged by the dedication of environmental organizations, but I recognize the challenges that must still be overcome to halt further destruction of the environment. Readers of this book will also identify many challenges of technology and within society for preserving Earth. Perhaps this book will give students inspiration to put their unique talents toward cleaning up the environment.

Acknowledgments

I would like to thank a group of people who made this book possible. Appreciation goes to Bobbi McCutcheon, who helped turn my ideas into clear, straightforward illustrations, and Elizabeth Oakes, for providing wonderful photographs that recount the story of environmental medicine. My thanks also go to Marilyn Makepeace, Jacqueline Ladrech, and Jodie Rhodes for their tireless encouragement and support. I thank Melanie Piazza, director of Animal Care, and the staff at WildCare, San Rafael, California, for information on animal rehabilitation. I appreciate the information provided by the Marine Mammal Center in Sausalito, California, for details on marine mammal rescue and release. Finally, I thank Frank Darmstadt, executive editor, and the editorial staff at Facts On File.

Introduction

One of the most troubling aspects of pollution in this age is its pervasiveness—there are few places left where a person or an animal can live without being exposed to pollution. Toxic substances travel great distances through the air, and many of these substances eventually fall from the atmosphere onto growing crops and open waters. Meat- and milk-producing animals ingest pollutants, produce growers spray large amounts of toxic pesticides onto food intended for people's dinner tables, and both animals and humans ingest small amounts of unhealthy compounds in water. As a result human physiology today may differ quite a bit from the physiology of people living in rural areas 200 years ago. Scientists now detect an array of toxic compounds (compounds that harm the body) in the blood and tissue of almost every person from whom they collect a sample. Surely wildlife suffers the same problem.

Environmental medicine developed into an important specialty within human and veterinary medicine when environmental science began uncovering increasing levels of pollution in the environment. This new medical specialty focused on the possible harm being done to health by exposure to these dangerous chemicals. In the past 30 years a large store of knowledge has grown on the health effects of specific toxic chemical classes. The medical community now understands that different *toxins* exert different effects on every system of the body's metabolism. A growing body of evidence has therefore been forming on the effects of heavy metals versus organic solvents, pesticides versus microscopic particles. Doctors have additionally learned more about toxic effects on organs, tissues, or cells.

Pollution: Treating Environmental Toxins covers the current knowledge in environmental medicine. This is an evolving science, so many questions remain concerning the way toxins disrupt body functions and the modes of developing disease. But this field is growing quickly in the

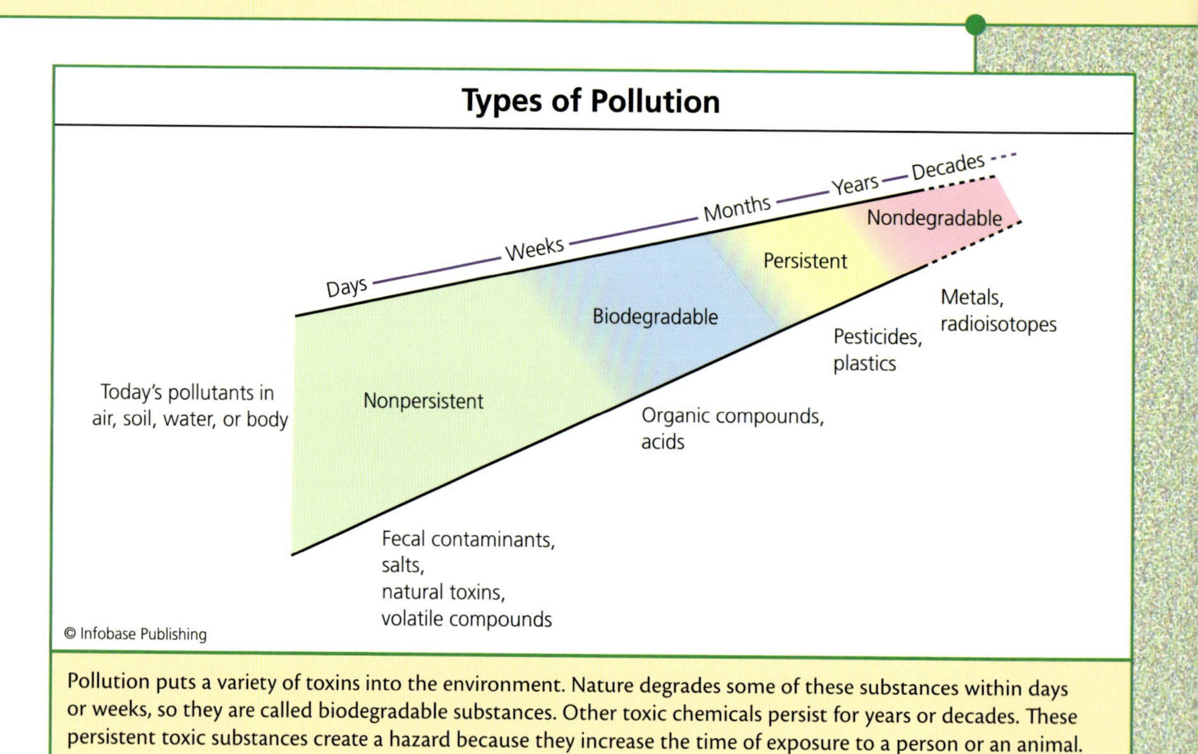

Pollution puts a variety of toxins into the environment. Nature degrades some of these substances within days or weeks, so they are called biodegradable substances. Other toxic chemicals persist for years or decades. These persistent toxic substances create a hazard because they increase the time of exposure to a person or an animal.

disciplines discussed in this book. Chapter 1 provides an overview of environmental medicine: how it began; the special interest areas within this field; the relationship between different types of people and their chance of being made ill from a toxin; and the international aspects of environmental medicine.

Chapter 2 discusses one of the foundation sciences used in environmental medicine: *epidemiology,* the study of disease outbreaks and spread in populations. The chapter highlights how statistics play an important role in finding disease patterns in very large populations. This chapter describes how proper diagnosis supports epidemiology and provides an overview of the current tools used for diagnosing disease. *Toxicology* studies also offer critical information on how toxins harm an individual or threaten certain groups within a larger population, so the chapter describes the science of toxicology and defines different types of toxicities, exposures, the accumulation of toxins in the body, and the chemical *persistence* of toxins in the body. Finally, chapter 2 discusses an area of great concern in public health: the relationship between pollution and cancer.

Chapter 3 focuses on specific toxin categories. It provides details on exposure, toxin sources, the ways in which people become contaminated with toxins, and how these chemicals affect the normal functions of organs, tissues, cells, and genetic factors. The chapter pays special attention to pesticides, organic compounds and solvents, and heavy metals, plus toxins that are emerging as health issues, such as endocrine disrupters and plastics. Chapter 3 concludes with sections on how the body detoxifies chemicals, the therapies used in environmental medicine to help natural *detoxification,* and preventive measures for avoiding exposure to toxins.

The next chapter delves into the subject of air pollution in regard to its health effects. Air quality, *greenhouse gases,* and carbon dioxide make up a major portion of all discussions on air pollution. This chapter also includes special types of hazards in the air such as electromagnetic fields, allergies, and excess noise. Chapter 5 follows the same theme by discussing the hazards in food and water. Water quality has become a critical problem in more and more areas of the world, both from water losses and increased pollution of drinking water. Though industrialized nations generally enjoy a clean, never-ending supply of drinking water, other parts of the world are approaching water crisis. Contaminated water and contaminated food has forced many societies to make choices between malnutrition and consuming toxins. Obviously, these societies cannot last long if left with no alternatives. This chapter puts emphasis on the global approaches to improving food and water quality and discusses how these problems have grown to global proportions.

Chapter 6 discusses the human populations that are of higher-than-normal risk for illness from environmental toxins. In fact, risk is one of the most important aspects of general medicine as well as environmental medicine, so this chapter puts extra emphasis on the concept of risk. The chapter also discusses the meaning of *demographics,* the effect of age on health, and the subpopulations in society that have high risks of exposure and illness from toxins.

This book closes with a chapter devoted to veterinary medicine as regards environmental sicknesses and injuries. Environmental toxins attack animal tissue, especially that of mammals, in a similar fashion as they damage human tissue. Chapter 7 deals with the threats put on wildlife by pollution entering their habitats. It also covers the new treatments being developed on captive animals in zoos and in rehabilitation centers that can be used to save wildlife when exposed to toxic levels of pollutants.

Environmental medicine in humans and in wildlife may soon grow into a critically important aspect of medicine if the world's oceans, land, and skies continue to fill with unnatural and hazardous substances. This medical discipline is therefore an essential part of environmental science and it may be more important with each passing day.

The Emergence of Environmental Medicine

Environmental medicine encompasses the *diagnosis* and treatment of illnesses caused by environmental toxins in air, water, or food. A toxin is a substance of chemical or biological origin that damages living tissue. Environmental toxins have caused growing concern since the 1980s because of repeated spills and leaks into the environment that have affected community health. The field of environmental medicine studies the health effects of chemicals released from industrial processes, vehicles, agricultural practices, and consumer products. Many of these chemicals have been released into the environment intentionally, but others have entered the environment because of accidents, such as oceangoing oil tanker spills. Environmental medicine also includes biological toxins that come from microbes or plants. The rise of diseases from biological toxins has been attributed to many factors that define the way the world works today: global travel, transoceanic shipping, changing immigration patterns, and food and water shortages. Climate change, too, may be considered the overriding cause of many new or reemerging *infectious diseases*.

The most troubling aspect of environmental toxins resides in their action in the body, which can take years or decades to become apparent as an illness. Industrial spills that occurred decades ago may only now show up in the form of cancer, genetic disorders, or other as yet undiagnosed ailments. Some environmentally caused ailments afflict the population in a more insidious manner. For example, the phenomenon known as *chronic sublethal poisoning* is the continual long-term exposure to a toxin that over time causes damage to the body, even if the toxin does not cause death. Advances in environmental medicine may someday show that almost everybody has some environmental toxins in their bodies that remain there for long periods.

As environmental chemists learn more about the pollutants hidden in soil and water and invisible in the air, environmental medicine will expand its list of suspected health dangers. This chapter covers the current status of human environmental medicine. It explains pertinent areas of disease diagnosis and treatment and then focuses on the specific challenges of finding solutions for illnesses caused by unknown agents in the environment. The chapter emphasizes the global view of environmental health. Just as endangered plant and animal species possess biodiversity hotspots, environmental medicine can be said to contain health hotspots, places where people have an increased *risk* of becoming sick from the substances in their environment. This chapter also explains key areas in environmental medicine today and future trends for how medicine will tackle new illnesses in the future. Considering the amount of pollution in many parts of the world, those future illnesses may not be far off.

THE GROWTH OF ENVIRONMENTAL MEDICINE

Environmental medicine mirrors general medicine by the steps that take place when a person feels sick: (a) the ill person sees a doctor; (b) the patient describes symptoms to the doctor; (c) the doctor draws on experience and laboratory test results to make the best possible attempt at identifying the illness; (d) the doctor prescribes a treatment for the illness; and then (e) the doctor follows the success of the treatment by monitoring the patient's symptoms. If the symptoms disappear, the identification and the treatment of the disease are presumed to have been correct. This speculation as to the cause of the symptoms, that is, the identification of an unknown disease, is a diagnosis. A final follow-up step involves the patient and doctor working together to devise a plan so that the illness never returns. This concise description of medicine seems straightforward, but the five steps seldom fall into place in the smooth fashion everyone would wish. Environmental medicine includes many challenges, the main ones being the following: a sick person does not see a doctor; a patient does not describe the symptoms completely or accurately; the doctor makes an incorrect diagnosis; the patient or doctor does not have access to the required treatments; or the patient does not comply with the treatment.

The problems listed here are real and common occurrences in global medicine. People in some parts of the world do not have the availability of medical care they need, but even in developed countries, doctors' best intentions can break down at any point in the process. Environmental medicine also must overcome the challenge presented by toxins that are hidden in the environment, undetected, but can cause illness many years after exposure. Additional common terms used in environmental medicine are listed in the following table.

Basic Terms in Environmental Medicine

Term	Definition
disease	condition defined by a patient's health complaints, a doctor's observations, and clinical test results
injury	physical damage to some part of the body; trauma
symptom	subjective evidence of disease, such as pain
sign	objective indication of disease that can be seen, such as *jaundice*
syndrome	group of symptoms, signs, laboratory findings, or physiological problems that are linked to a known disorder or disease
chronic	medical condition of long duration or course
acute	medical condition of rapid onset, severe symptoms, and a short duration or course
exposure	amount of hazardous substance received by a part of the body or the entire body
dose	measurable amount of a substance put into a body

(continues)

BASIC TERMS IN ENVIRONMENTAL MEDICINE
(continued)

TERM	DEFINITION
pathogen	disease-causing microorganism, or microbe
prophylaxis	a plan for preventing disease
surveillance	disease-monitoring process for studying epidemics, outbreaks, or health risks
toxin	poisonous substance of plant or animal origin or a poisonous chemical
toxicity	degree to which a substance is poisonous
poisoning	illness produced by taking a toxin into the body
occupational health	medical status related to a person's work
environmental health	medical status related to substances in a person's surroundings
public health	processes for preventing disease and improving the health of a community
toxic chemical	chemical that causes temporary or permanent harm or death in humans or animals
hazardous chemical	chemical that may potentially cause harm to health because it is ignitable, reactive, corrosive, or toxic

Environmental medicine's roots originate with the birth of epidemiology, which is the study of health and illness distributed across a population. The first case in environmental medicine occurred with a sudden outbreak of the bacterial disease cholera in England, in London's Soho district in 1854. Local doctor John Snow noticed the symptoms of his sick patients always seemed to be associated with their digestive tracts. Without being sure of much else, he proposed to other doctors that patients were ingesting something—food or water—and this unknown agent caused the disease. Snow's idea ran counter to the prevailing belief at the time that diseases mysteriously floated through the air, called "bad air." As sick people poured into every medical office in Soho, Snow visited their homes, learned about their routines, and traced the water each afflicted and each healthy family had been using. His detective work identified a single water pump that almost all of

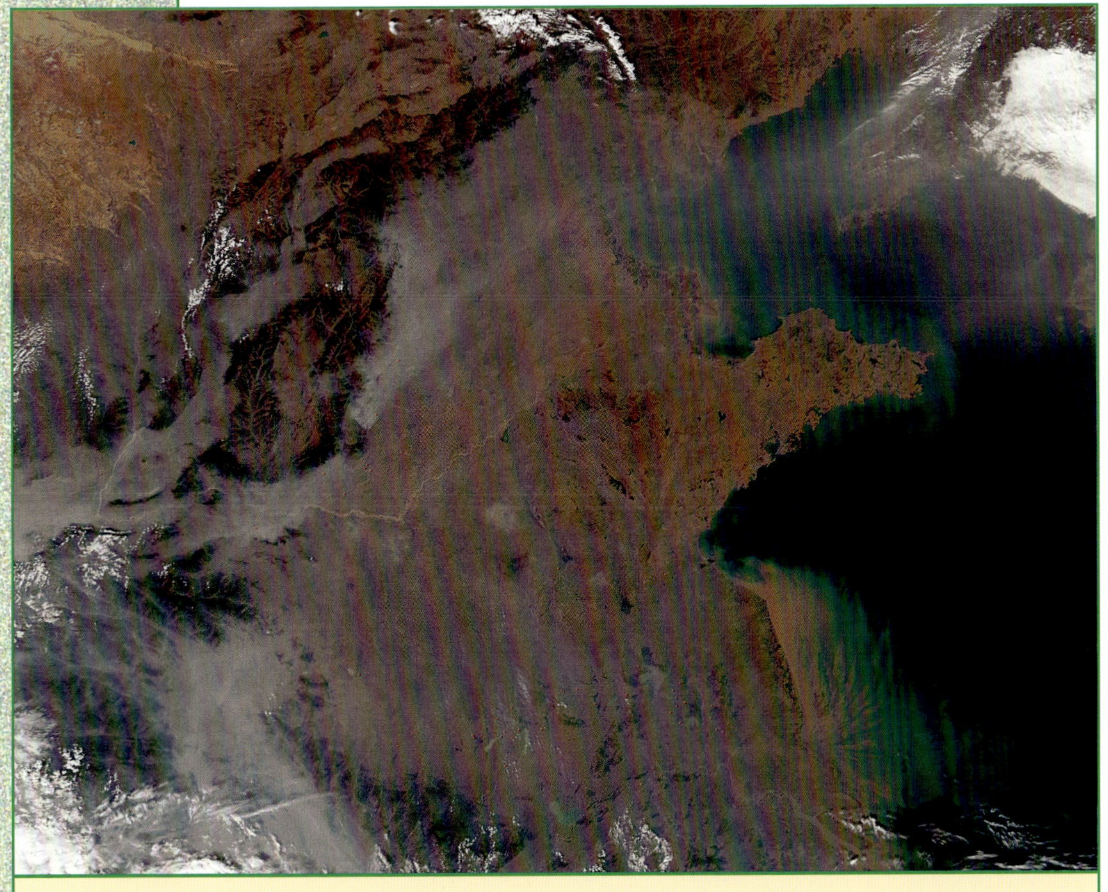

This satellite photo taken over eastern China in 2002 shows the diverse effects of pollution on the environment. Dozens of fires (small red dots) create smoke that clouds the skies and funnels through valleys. Much of the air pollution blows east toward Korea and the Pacific Ocean. Meanwhile, the Yangtze (Chang) River, at lower right, deposits brownish, sediment-laden waters into the Yellow Sea. *(NASA)*

the stricken had in common. Snow immediately had the pump's handle dismantled so no one could reach the underground water, and before long Soho's cholera epidemic began to fade. Snow wrote of the disease-causing agent, "The poison consists probably of organized particles, extremely small no doubt . . ." He went on to describe what today would clearly be an environmental medical concern. Snow felt he had gathered ". . . very strong evidence of the powerful influence which the drinking water containing the sewage of a town exerts on the spread of cholera when that disease is present." In its straightforward way, Snow's experience described how to find the source of environmental illness.

Like London's cholera epidemic, most of the environmental toxins before the 1900s came from microbes or parasites rather than chemicals.

Toxic chemicals have harmed wildlife as much if not more than they have harmed humans. This peregrine falcon chick is fortunate to have been born in Wisconsin years after the U.S. ban on pesticides such as DDT, which almost eliminated the peregrine population. The peregrine falcon was the first animal put on the endangered species list in 1970. By 1999 peregrine populations had recovered, and the bird was removed from the list. *(University of Wisconsin–Stevens Point, Wildlife Society chapter)*

The list of new chemicals entering the environment grew as new industries emerged in the 20th century. The United States began taking the world lead as an industrialized nation at the turn of the century. In 1870 the United States accounted for 23 percent of the world's manufacturing; by 1913 it performed more than 35 percent. As the industrial revolution unfolded, pollution trickled into the environment while natural resources began to disappear. By the middle of the century, pollution no longer trickled: in some places it poured.

At the end of World War II in 1945, synthetic chemicals made up an increasingly larger proportion of the pollutants entering the environment. Plastics, polychlorinated biphenyls (PCBs), and a new pesticide called dichlorodiphenyltrichloroethane, or DDT, began accumulating in soils and waters. In 1977 a heavy rainfall leached chemicals from buried, corroded drums holding 21,000 tons (19,051 metric tons) of industrial chemicals in an area outside Niagara Falls, New York. The wastes seeped into homes, schools, and playgrounds. The area, known as Love Canal, became the United States' most famous environmental health disaster. Residents of the area who had been exposed to the chemicals developed cancers and other serious illness, and suffered miscarriages and an increased incidence of birth defects. But Love Canal proved to be the tip of an iceberg of environmental contamination around the world. The following table lists the most serious environmental accidents.

As the table shows, environmental accidents declined as the new environmental laws enacted in the 1970s gradually took shape and gained strength in the following years. Accidents continue to happen, but U.S. laws began having a positive effect. Worldwide, however, illnesses caused by pollution remain serious in some places.

Environmentally triggered illnesses (ETI) make up one aspect of environmental medicine. An ETI occurs when a substance in the environment stresses the body's normal functions. Diagnosis of an ETI takes into consideration a patient's health history in addition to the type of environmental toxin, the amount of exposure, and the time duration of the exposure. ETI medicine has been taught in medical schools only since the 1970s. Even today, family physicians may not have the resources to investigate a patient's background to explore a potential environmental *toxicity.*

Another challenge in environmental medicine resides in the legal issues that revolve around an environmental accident. Industrial pol-

Place	Year	Event
United States		
Donora, Pennsylvania	1948	air pollution from local factory kills nearly 100 people
Cuyahoga River, Ohio	1969	oil slick catches fire on the river
Three Mile Island, Pennsylvania	1979	nuclear reactor accident releases radiation into neighboring community
Love Canal, New York	1980	industrial canal leaks toxic chemicals into homes, schools, and other buildings
Mount Saint Helens, Washington	1980	volcanic eruption releases gases, chemicals, and particles into the atmosphere
Exxon Valdez, Alaska	1989	oil tanker spills more than 1 million barrels of crude oil, causing environmental disaster in Prince William Sound
Hurricane Katrina, Louisiana, Mississippi, Alabama, Florida	2005	one of the nation's worst natural disasters, flooding causes toxic chemicals and pathogens to pollute the Gulf Coast and kills more than 1,800 people and decimates wildlife

lution cases can be tied up in court for decades after the incident, and indeed, victims have died with an affliction while legal wrangling goes on. Other parts of the world confront greater hurdles in preventing environmental illness. For instance, impoverished regions likely suffer from malnutrition, lack of water, and lack of medical care. These conditions can open the door to environmental toxicities or make existing toxin poisoning worse.

PLACE	YEAR	EVENT
Global		
Minamata disease, Japan	1932–1968	mercury contamination affects more than 3,000 victims and causes severe birth defects
Seveso, Italy	1976	chemical spill pollutes more than seven square miles (18 km^2) and became highest known exposure to a dioxin
Amoco Cadiz, France	1978	tanker grounding spills 1.6 million barrels of crude oil into Atlantic Ocean off Brittany
Bhopal, India	1984	toxic gas leak kills several thousand people
Chernobyl, Ukraine	1986	nuclear power plant accident causes an explosion
Sandoz, Switzerland	1986	chemical spill pollutes miles of the Rhine River
Gulf War, Persian Gulf	1991	burning of more than 600 oil wells pollute air, land, and coasts
Baia Mare, Romania	2000	55–110 tons (50–100 metric tons) of cyanide and heavy metals spill into the Danube and other rivers

THE GLOBAL STATUS OF ENVIRONMENTAL HEALTH

Global environmental health encompasses all the major health risks that could occur when people interact with their environment anywhere in the world. The following list gives examples of topics in environmental health:

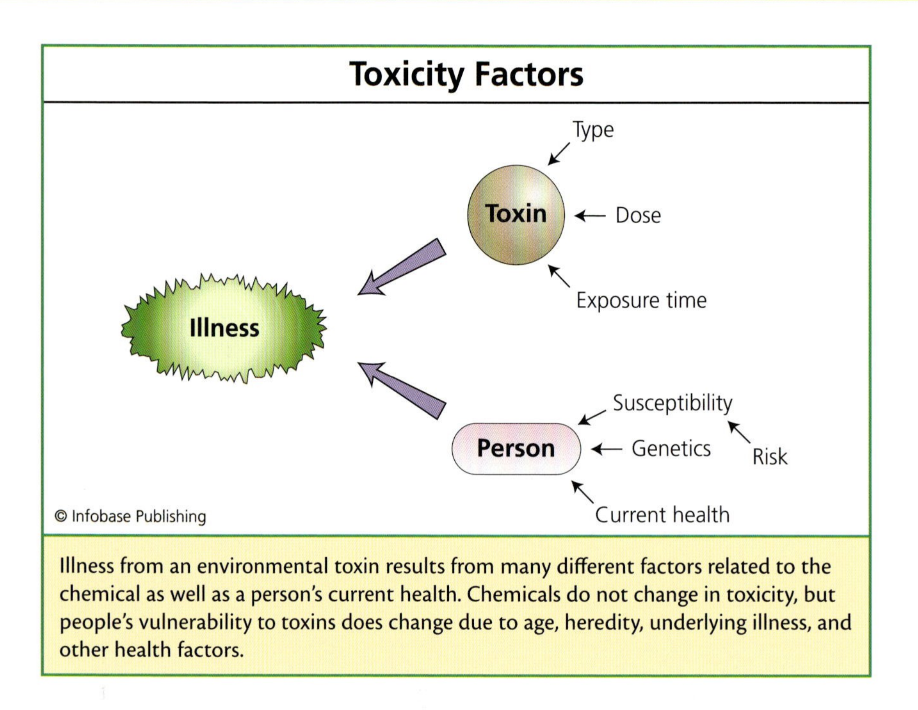

Illness from an environmental toxin results from many different factors related to the chemical as well as a person's current health. Chemicals do not change in toxicity, but people's vulnerability to toxins does change due to age, heredity, underlying illness, and other health factors.

- a pathogen that spreads in a region's drinking water source
- pesticide contamination of field crops
- mercury in dangerous levels in ocean-harvested fish
- algae toxins that make beaches unsafe for swimming
- an oil slick that coats and kills thousands of waterfowl

Human and wildlife both fall victim to environmental toxins, but wildlife do not have the luxury of being warned not to drink contaminated water or eat poisoned plants. Children also would have little reason to know the hazards in their environment, and worldwide, children's health is a major issue in environmental medicine. Children are especially susceptible to toxins because their bodies are still growing and developing, and because pound for pound children eat, drink, and breathe more than adults. The Centers for Disease Control and Prevention has stated that "about one-quarter of global disease is caused by avoidable environmental exposures; for children under the age of five, the figure is more than one-third." Malnourished children run greater risks of harm from toxins because their bodies are already in an unhealthy state.

The U.S. Department of Health and Human Services has stated, "Actions in every country affect the environment and influence events around the world." Today global travel and international economies can move hazardous substances halfway around the world within 48 hours. Doctors in general do not know the extent of illness that comes from hazards far away, but studies on animals living in the Antarctic have detected pesticides in their tissue, meaning the chemicals traveled halfway around the world.

People and animals withstand extremes of temperature for short periods of time, but any living thing that is stressed or in a diseased state has a harder time adjusting to environmental extremes. Changes in average environmental temperatures affect biota on land and in aquatic and *riparian* environments. Global warming causes additional concerns to world health in the following ways: disease transmission by *vectors* such as mosquitoes; pathogen transmission through water; desertification causing food shortages; increasing *water stress* due to drought; and the possibility of stronger, more frequent tropical storms. All of these events either help spread disease or make people or animals more vulnerable to disease.

The World Health Organization (WHO), discussed in the sidebar on page 13, covers most aspects of global health today, including travel that spreads infectious diseases, toxins, pollutants, and *invasive species* much faster and farther than they had spread in any other time in history. There are four global health hazards monitored by the WHO: food-borne disease; chemical pollution; radioactive contamination; and environmental disasters that disperse disease and toxins. Chemical and microbial toxins can contaminate food supplies, and the increased international trade in foods, grains, oils, and processed products move these health hazards from country to country.

Massive chemical or radioactive contaminations of air, water, and earth draw people's attention more than slow and steady contamination, such as that in certain water sources. Large accidents highlight the threats to not only humans, but also wildlife and aquatic life. Nuclear accidents such as the 1986 explosion in Chernobyl in eastern Europe add to many people's fear that nuclear power looms as the world's big-

Millions of people have become needlessly harmed by environmental toxins. Some regions do not have strict controls on chemicals released into the environment. Political conditions may also make a population vulnerable to toxins; war, starvation, lack of water, and an absence of medical care increase the health threats from toxins. This mother in a refugee camp in North Darfur in Africa holds her malnourished 27-month-old child. *(USAID)*

gest environmental threat. A few months after the Chernobyl event, Andronik M. Petrosyants spoke for a Soviet committee on peaceful uses of atomic energy: "The accident at the Chernobyl atomic-power station badly affected Soviet atomic-power engineering, and will undoubtedly have an effect on the world's atomic-power industry as a whole." In contrast to this concern over industry's future, the Soviet government was slow to care for residents that had received large exposures to radioactivity.

Like chemical accidents symbolized by Chernobyl, natural disasters such as Hurricane Katrina in the southeastern United States in 2005 also serve as a large one-time source of toxins. The hurricane wiped out hundreds of thousands of animals and eradicated habitat, spread pollution, and tainted the water supply of millions of people. The environment does not recover quickly from such cataclysmic events, and governments may be the only feasible way to deliver medical care and enforce precautions for the future.

ENVIRONMENTAL HEALTH SPECIALTIES

Occupational health and environmental health are related in that they both focus on health hazards in a person's surroundings. Occupational health relates to the diagnosis, treatment, and prevention of illness or injury resulting from a person's work. Occupational health overlaps with

The World Health Organization

The World Health Organization (WHO) grew out of plans made at the United Nations first meeting in 1945, and it began operations in 1948. WHO's acting president, Andrija Štampar of Yugoslavia, recapped the events: "Over two years have elapsed since the representatives of more than sixty nations assembled in New York at the International Health Conference and decided—on the suggestion of the representatives of China and Brazil—to establish the World Health Organization." President of the Swiss delegation, M. Etter, added, "In comparison with past programs of international cooperation in the field of public health, this . . . marks a great step forward, indeed almost a revolution. . . . Its [WHO's] effort will not be confined to fighting the dangers which threaten the health of peoples: they will more especially be directed towards developing well-being and health in general, embracing the whole nature of man, physical and spiritual . . ." The WHO has now grown to almost 200 members that continue to follow this philosophy.

The WHO contains representatives from almost all members of the United Nations and comprises the following six regions (regional offices in parentheses): Africa (Cité du Djoué, Congo), the Americas (Washington, D.C.), Southeast Asia (New Delhi, India), Europe (Copenhagen, Denmark), Eastern Mediterranean (Cairo, Egypt), and the Western Pacific (Manila, Philippines).

The organization covers more than 100 specific health topics, including environmental health. The major programs study issues in indoor and outdoor air pollution, chemical exposure, electromagnetic fields, *ultraviolet radiation* and *ionizing radiation*, water sanitation, and environmental health issues in children. In addition to publications in these areas, the WHO provides resources to physicians and the public in the areas of international travel, trends in environmental threats, and the future impact of climate change on human health. Perhaps the WHO's greatest contributions come from two facets of its ongoing work: providing guidance and direction to research in global public health, and helping diverse countries cooperate on health issues that affect them all.

environmental health when workers become exposed to environmental hazards. The following events might be considered occupational, environmental, or both: lung cancer caused by breathing asbestos fibers; black lung disease acquired by coalminers; nervous disorders experienced by people fishing algae-infested waters; or skin rashes on workers assigned to clean a hazardous waste site. Environmental health may be distinguished from these examples because it relates only to illness caused by a hazard in a person's environment and not solely to hazards at a job.

Environmental health contains a wide range of specialized areas connected with environmental toxins, disease diagnosis, treatment, and other doctor-patient relationships. The main specialties in environmental health are the following:

- patient education—teaching people about the hazards in their environment and how to reduce the risk of exposure to the hazards
- immunotherapy—involves mainly the vaccination and prevention programs used for new or reemerging infectious diseases
- neurology—an important part of environmental medicine because many chemical pollutants cause nerve disorders
- detoxification therapy—the science of neutralizing a toxin that has contaminated a patient's blood or tissue
- environmental controls—mechanisms for reducing exposure to toxins through the use of air and water filters, nutritional supplements, or protective clothing
- surgery and treatment—any corrective actions to toxin-caused disease

Since ETIs can affect any organ in the body, environmental medicine contains the same specialties that are part of western medicine: cardiovascular disorders; eye/ear/nose/throat disorders; pulmonary problems; endocrine dysfunction; gastrointestinal disorders; hematologic problems; genitourinary disorders; neurological injuries; musculoskeletal malfunctions; behavioral and psychiatric disorders; skin abnormalities; and cancer. Examples from each of these specialties range from well-defined diseases, such as asthma or cancer, to very general syndromes that may be difficult to diagnose, such as attention deficit disorder, anxiety, sexual dysfunction, or eating disorders.

The following table gives examples of three classifications of diseased states that may be the result of an environmental hazard. Top-level conditions consist of defined symptoms and the availability of medical tests that enable a straightforward diagnosis. Second-level conditions may be more difficult to diagnose or treat or may present difficulty in identifying their cause. Third-level conditions contain chronic problems that may be the result of an underlying illness or injury.

The study of ETIs involves the toxicology of an illness caused by environmental factors. Environmental toxicology is a relatively new topic in medicine that concerns the harmful effects of chemicals, physical agents, or biological things on the health of living things, including plants, animals, fish, and humans. The toxic substances that cause the most harm to *biota* can be classified various ways, but in general, toxicology involves the following categories of hazardous substances: metals, chemicals, organic solvents, pesticides, gases and airborne particles, and biological agents. Several substances fit into more than one category, such as compounds of mercury (a metal) that might be emitted from

Examples of Conditions that Result from Environmental Factors

Category	Conditions
top level	allergy, asthma, cancer, irritable bowel syndrome
second level	attention deficit disorder, chronic bronchitis, chronic fatigue, dermatitis, migraine headaches, sinusitis
third level	anxiety, chronic gastritis, depression, eating disorders, frequent colds, hypertension, irritability, memory loss, Parkinson's disease, sexual dysfunction, thyroid dysfunction, vertigo

smokestacks as an airborne particle or contaminate the soil. Other classifications in toxicology may simply assign categories to the source of the toxin, such as air, water, soil, or food toxins.

The toxicology of any pollutant may be analyzed by two scientific specialties: *toxicokinetics* or *toxicogenomics.* Toxicokinetics is the study of movements of toxic substances throughout the body and the relationship between dose and harm to the body. Specifically, toxicokinetics focuses on four main aspects of environmental toxins in a living thing: absorption, distribution, metabolism, and excretion. The four routes by which toxins usually enter the body are gastrointestinal (by ingestion), pulmonary (inhaling into the lungs), percutaneous (through the skin), or ocular (through the conjunctiva of the eyes).

Toxicogenomics is the study of how a living thing's genes respond to environmental stress or toxins. This science is important for assessing how wildlife responds to environmental changes and their means of adapting to change. Toxicogenomics is described in more detail in the sidebar on this topic on page 18.

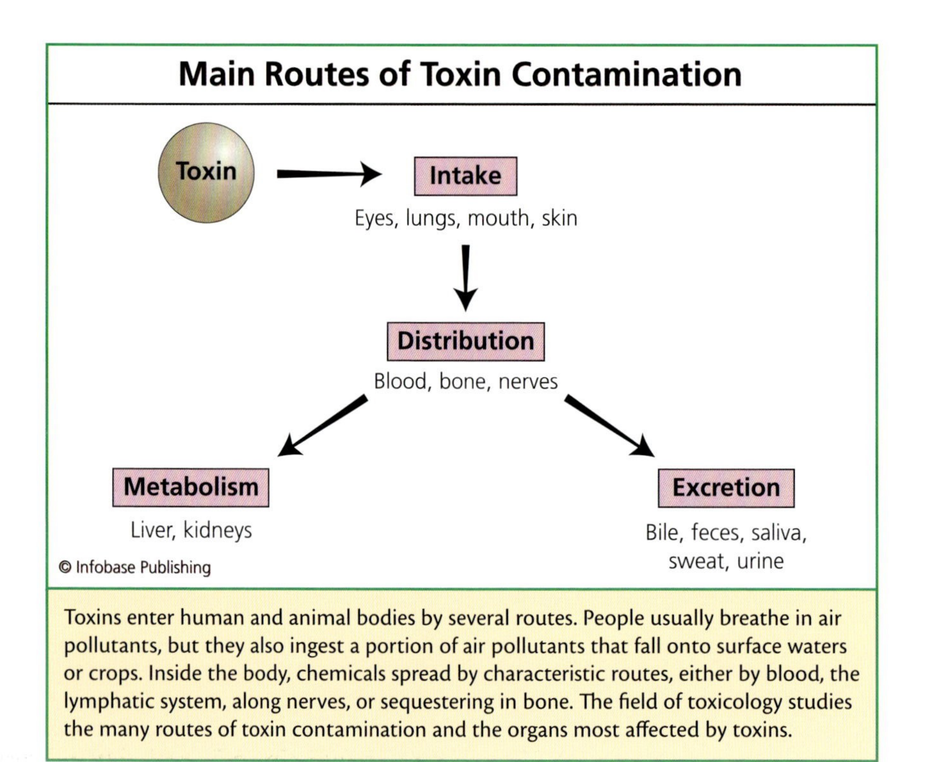

Toxins enter human and animal bodies by several routes. People usually breathe in air pollutants, but they also ingest a portion of air pollutants that fall onto surface waters or crops. Inside the body, chemicals spread by characteristic routes, either by blood, the lymphatic system, along nerves, or sequestering in bone. The field of toxicology studies the many routes of toxin contamination and the organs most affected by toxins.

DEFINING ENVIRONMENTALLY TRIGGERED ILLNESSES

ETIs arise for various reasons in addition to the dose of a toxin and the period of time to which a person has been exposed to it. Most of this variation arises from the natural differences from person to person. Organisms differ in the following ways: the body's biochemical reactions; susceptibility; ability to adapt to a toxin; bipolarity (having only extreme positive or negative reactions to a stimulus); the spreading phenomenon (how easily the toxin disperses in the body); and the switch phenomenon (the body's ability to switch from severe toxic reaction to no reaction, or vice versa).

In 2005 a Canadian study determined the variety of environmental toxins in the bodies of volunteers, a process called *biomonitoring.* The study showed that even families with a seemingly healthy lifestyle and sound diet may have more than 35 different environmental chemicals in their blood and tissue. But knowing a chemical has contaminated a person is not the same as knowing what illness, if any, it causes. Vivian Maraghi, one of the study's subjects, said to Montreal's *Gazette* in 2008 that her 12-year-old son never gets sick, but she has suffered from migraine headaches for years. "I have had them since I was young," she said, "but it's hard to relate it to anything." ETIs have been difficult to diagnose, and their characteristics may require studies of many thousands of people to find a link between toxins and illness. Joe Schwartz, a chemist at McGill University in Montreal, Canada, elaborated on Maraghi's concern, saying, "The value of biomonitoring is going to be long-term. If we have a good baseline now, we get good data and then we check ten, twenty years down the road to see if there is any alteration in disease patterns for those people and then you look back to see of there is any link." Biomonitoring will provide valuable information on the connections between environment and disease, but this information will not come quickly.

Defining ETIs calls for observation plus statistics, as suggested by Schwartz's explanation. Biomonitoring helps form a picture of disease, especially in the two following aspects: (1) risk factors, meaning an individual's characteristics that make the person more or less likely to become ill, and (2) the demographics of the population being monitored. Demographics is the study of a population's characteristics in age,

TOXICOGENOMICS

Toxicogenomics defines the connection between three factors in environmental health: the toxin, a person's complete set of genes (called a *genome*), and the genetic response to a toxin-caused disease.

Organisms respond to their environment by altering the expression of genes, meaning the body's cells decide which genes are to become active in protein production and which genes are to be suppressed and temporarily turn off protein production. A large molecule called *messenger ribonucleic acid* (mRNA) receives the information contained in each gene and converts it to a code that tells *enzymes* how to build new protein. Proteins carry out most of the body's functions in the form of enzymes or as cell building blocks, often in combination with fats.

A toxic substance such as mercury may exert an effect on genes and so change the type and amount of proteins the body's cells produce. The altered proteins may leave the body vulnerable to disease, or these altered proteins may cause disease by affecting normal cell function. At the present time, doctors know of many illnesses causes by pollutants, but the exact manner in which the toxin makes cells' normal activities fail remains only partially understood.

Many toxins—metals and organic solvents are examples—behave somewhat alike inside the body, so researchers can gain general knowledge about more than one toxin at the same time. In the next few years, toxicogenomics will tackle the following questions in regard to the body's response to toxins:

- Do specific toxins produce unique changes in select genes?
- Do different cells in different tissues respond differently to the same toxin?
- Do different species have different, overlapping, or identical responses to toxins?
- How does age or current health status affect responses to a toxin?
- How do responses to chemical mixtures differ from the responses to one toxin?
- Can gene testing reveal a chronic low-level exposure to a toxin?
- Do individuals have certain genes that make them more or less susceptible to harm from toxins?

gender, ethnicity, and other features. Climate change has begun to change the demographics of plant and animal populations. The following sidebar, "Case Study: Climate Change and Infectious Diseases," explores how climate change may affect the environmental threats to people.

Gene Expression and Protein Production

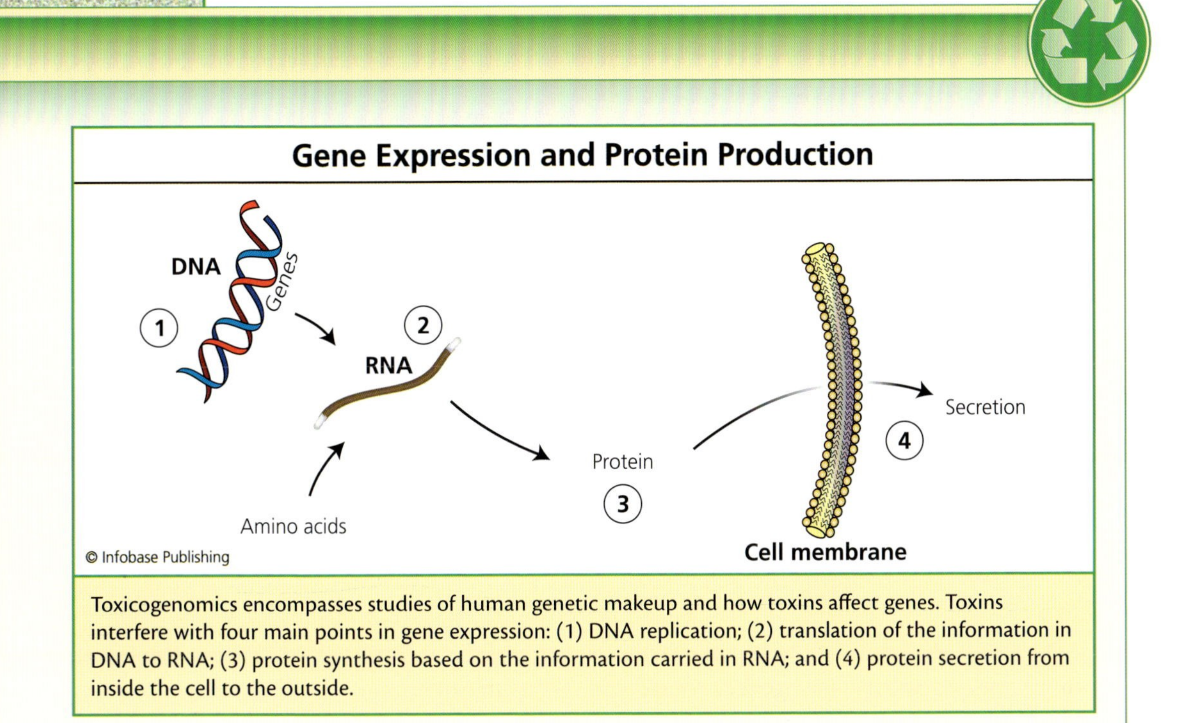

Toxicogenomics encompasses studies of human genetic makeup and how toxins affect genes. Toxins interfere with four main points in gene expression: (1) DNA replication; (2) translation of the information in DNA to RNA; (3) protein synthesis based on the information carried in RNA; and (4) protein secretion from inside the cell to the outside.

Sciences such as toxicogemonics must do studies on small groups of subjects (plants, animals, humans, etc.) and then relate the results to an entire population. This is because no technique can measure every individual in a population. But small subject groups may not always relate exactly to a larger population because in a large population no two people are identical. Variations arise from differences in individuals' genes. Multiple versions of a specific gene in a population are called *gene polymorphisms*, and these polymorphisms explain why one person exposed to a toxin for X number of years may develop cancer, but another person receiving the same dose for the same period remains healthy. Individuals who get sick from a toxin while other people do not are said to have a *genetic predisposition* to the illness. Scientists in toxicogenomics today concentrate on the phenomena of gene polymorphisms and genetic predisposition.

HEALTH RISK AND DEMOGRAPHICS

Risk relates to chances—in statistics this is called probability—of becoming injured or sick. A health risk is any characteristic belonging to an

Case Study: Climate Change and Infectious Diseases

Diseases have always found their way into human civilization. Though the plagues of the Middle Ages pointed out the need for decent sanitation, clean water, and proper waste disposal, even healthy communities deal with environmental threats. Global climate has been a factor in environmental hazards because it speeds the global movement of diseases and toxins.

The WHO has identified the following scenarios in which climate change, specifically global warming, and worldwide health relate to each other:

- increasing frequency of heat waves
- variable precipitation affecting freshwater supplies and waterborne diseases
- variable precipitation patterns with rising temperatures compromising food production, leading to increased malnutrition
- rising sea levels displace populations, putting stress on water and food supplies and sanitation
- lengthened disease transmission seasons and altered geographic ranges of disease
- emergence of new diseases and reemergence of diseases that were at one time controlled

Malaria carried by mosquitoes offers an example of a parasitic disease that may return to populations with global warming. Rising sea levels caused by climate change will lead to flooding, which in turn gives the *Anopheles* mosquito species more breeding grounds. Warmer average temperatures would also increase the mosquito's season. Global climate change can and will create other complex relationships between environments, human susceptibility to disease, and the transmission of infections.

individual that makes that person more susceptible to illness. The following list of health risks are known to make a person more vulnerable to illness than a healthy individual. Each of these groups represents an *at-risk group* or a high-risk group.

- infants and young children
- the elderly
- pregnant women
- people with diabetes
- *immunocompromised* people (such as transplant patients and AIDS patients)
- cancer patients
- people with a serious preexisting or chronic disease
- patients recovering from surgery
- people with chronic pain
- people with a family history of cancer or heart disease

Some of the groups listed here can raise the statistical health risk of an entire population if the high-risk group makes up most of the population. For instance, a nursing home contains a larger proportion of elderly people than the general population. An older subpopulation has a greater risk of illness from environmental toxins because it is a high-risk group.

The makeup of the population living in a nursing home, or a city, or an entire continent is called the demographics of the population. London physician John Snow used location demographics to determine that families living close to a contaminated water pump had a higher likelihood of contracting cholera than people living farther from the pump.

Socioeconomic factors help researchers determine the pattern of diseases. Wealthy regions have good medical care, a high level of education, balanced diets, and families likely to live far from environmental hazards. By contrast, impoverished regions lack these niceties and the general health may be poor, leaving residents open to new illnesses.

Demographics aid scientists find the root of an environmental problem, but success ultimately rests on hard investigation and a bit of luck in finding people that may have moved away from an unhealthy environ-

The wealth and medical care within a country influence the ability of the population to resist illness. These Sudanese women and children in a refugee camp in Sudan wait for therapeutic milk distributed by a hospital in the town of Iriba in Chad. Therapeutic milk contains ingredients added specifically to aid the health of those who are severely malnourished. *(UNICEF)*

ment. In the past few decades, a 100-mile area along the Mississippi River from Baton Rouge to New Orleans, has been nicknamed "Cancer Alley." (Industries prefer to call the area "Chemical Alley.") This region houses a dense population of industrial plants and the soil contains toxic metals, arsenic, pesticides, petroleum products, and an assortment of organic chemicals and solvents. Demographic studies have suggested that this region may have higher than normal cancer rates. *BBC News* provided an unbiased view of this question in a 2002 news report: "Those most often at risk are citizens living in small, low-income, predominantly African-American communities. Residents in this area suffer disproportionate exposure to the environmental hazards that come with living near chemical waste. Cases of rare cancers are reported in these communities in numbers far above the national average. For example, in the town of Gonzales, Louisiana, three cases of rhabdomyosarcoma, an extremely rare and devastating childhood cancer, were reported in a fourteen-month period. The U.S. national average of rhabdomyosarcoma is one child out of a million." Controversy can arise from places such as Cancer Alley because of the challenge in linking environment to disease. Johns Hopkins public health scientist Ellen Silbergeld told the *New Orleans Times-Picayune* in 2000, "This is not a sensitive science. There is just an absolute limit to what we can find out." For instance, testing for the safety of chemicals often takes place in laboratory animals, but animal data do not always equate with human biology.

Confusion about environmental data can lead to questions of whether health hazards truly exist. Physician Elizabeth Fontham of Louisiana State University pointed out to the *Times-Picayune* regarding Cancer Alley, "It's not like an infectious disease where you take a blood sample and there it is. Often the carcinogen [cancer-causing substance] is long gone." Simply put, both pollutants and people move around, making demographic studies difficult.

Toxin contamination in the body also complicates environmental medicine because a long time can occur between exposure and illness. Journalist David Ewing Duncan wrote in 2006 for *National Geographic* magazine, ". . . I grew up in northeastern Kansas, a few miles outside Kansas City. There I spent countless hot, muggy summer days playing in a dump near the Kansas River. Situated on a high limestone bluff above the fast brown water lined by cottonwoods and railroad tracks, the dump was a mother lode of old bottles, broken machines, steering wheels, and other items only boys can fully appreciate. This was the late 1960s, and my friends and I had no way of knowing that this dump would later be declared an EPA [U.S. Environmental Protection Agency] superfund site, on the national priority list for hazardous places. It turned out that for years, companies and individuals in this corner of Johnson County had dumped thousands of pounds of material contaminated with toxic chemicals here." Many people in the United States and the world have the same experiences of having once lived very close to pollution even though they live in cleaner places now.

One of the hundreds of chemicals that infiltrated people's lives during the period journalist Duncan described was a pesticide called dichlorodiphenyltrichloroethane, better known as DDT and highlighted in the "DDT" sidebar on page 24.

BIOMAGNIFICATION

Biomagnification is the process in which a chemical becomes increasingly concentrated as it moves in plant or animal tissues up a food chain. The reason chemicals increase in concentration, or magnify, comes from the nature of food chains. At the bottom of an aquatic chain, for example, millions of tiny plankton bodies may contain one ten-thousandth of a part per trillion (0.0001 ppt) of mercury. Small fish consume millions of plankton and accumulate the mercury in their tissue, perhaps as high as

DDT

DDT was discovered in the late 1800s, but not until 1939 did the Swiss company Geigy Pharmaceutical study DDT for its commercial uses. DDT belongs to a class of pesticides known as insecticides, compounds that kill insects. By the mid-1940s crews trying to control mosquito breeding and lice infestations found DDT to be a powerful weapon; the WHO estimates DDT saved several million lives from malaria and typhus, carried by mosquitoes and lice, respectively. Farmers used DDT to protect crops and public health officials continued to rely on DDT for disease control; in 1959 alone, they applied 40,000 tons (36,290 metric tons) of DDT to the land.

DDT kills insects by destroying their nervous system, but the chemical also dissolves in fat and so stays in the body of any animal that ingests it. DDT has a half-life of 15 years, meaning it takes that long for only one-half of the chemical to degrade. During this time the chemical has a chance to accumulate to higher concentrations in tissue with each step up a food chain in a process called *biomagnification.* Animals at the top of each DDT-contaminated food chain receive the highest dose of DDT.

In 1948 chemist Paul Muller was awarded the Nobel Prize in medicine and physiology for determining the pesticide activity of DDT as it contributes to global health. But in 1962 a nature writer and zoologist with extraordinary powers of observation, Rachel Carson, wrote the book *Silent Spring,* and in it she traced the possible routes of DDT from manufacturing plants into the environment, and then into the bodies of wildlife. In her book, Carson warned, "DDT is now so universally used that in most minds the product takes on the harmless aspect of the familiar." Carson had succinctly described the movement of a chemical through a food chain. "One of the most sinister features of DDT and related chemicals are the way they are passed from one organ-

0.01 ppt. Larger fish then consume small fish, and since each small fish has accumulated the mercury, the larger fish receive a large dose. A large fish such as tuna may contain two parts per million (ppm) of mercury in its tissue, for instance. At the top of the food chain, humans can accumulate dangerous levels of mercury by eating a steady diet of contaminated tuna. In some instances, animals at the top of food chains may be severely injured or killed by the high level of contamination in their food sources.

Bioaccumulation occurs when chemicals stay in the body and become stored there faster than the body can degrade them. This process acts as a survival mechanism for animals for storing nutrients, but animals store

ism to another through all the links of the food chains. . . . Hay, containing 7 or 8 parts per million, may be fed to cows. The DDT will turn up in the milk in the amount of about 3 parts per million, but in butter made from this milk the concentration may run to 65 parts per million." Other pesticides are now known to follow this same biomagnification pattern.

In wildlife—as Carson had feared—scientists began to see declines in animals that were at the top of food chains, such as bald eagles and peregrine falcons. DDT had been accumulating in the birds' tissue and caused eggshells to be brittle. The weakened eggs broke before each embryo could develop; eagle and falcon numbers plummeted. On December 31, 1972, the EPA released this news to the public: "The general use of the pesticide DDT will no longer be legal in the United States after today, ending nearly three decades of application during which time the once-popular chemical was used to control insect pests on crop and forest lands, around homes and gardens, and for industrial and commercial purposes." Though the U.S. use of DDT stopped, other countries continue to depend on it to control mosquitoes.

Even some environmental groups have come to realize the worth of DDT in human health. Sierra Club director Ed Hopkins told the Public Broadcasting System (PBS) in 2006, "Malaria kills millions of people and where there are no other alternatives to indoor use of DDT, and where that use will be well-monitored and controlled, we support it. Reluctantly, we do support it." The world is left to decide between DDT and infectious disease. The WHO recommends the use of DDT for blocking the transmission of malaria, but this organization also calls for alternative practices that may reduce the world's use of this known carcinogen and threat to wildlife health.

environmental toxins by the same mechanism, and to lethal levels. Large animals with long life spans and substantial fat stores are at greater risk from toxin bioaccumulation than small animals with rapid metabolism.

CONCLUSION

Environmental medicine specializes in diagnosing and treating illnesses causes by toxins in the environment. This area of medicine has grown with the increasing incidence of chemical spills. The nature of environmental toxins presents challenges in this field of medicine. First, environmental toxins exist in the environment at unknown levels, so it is difficult

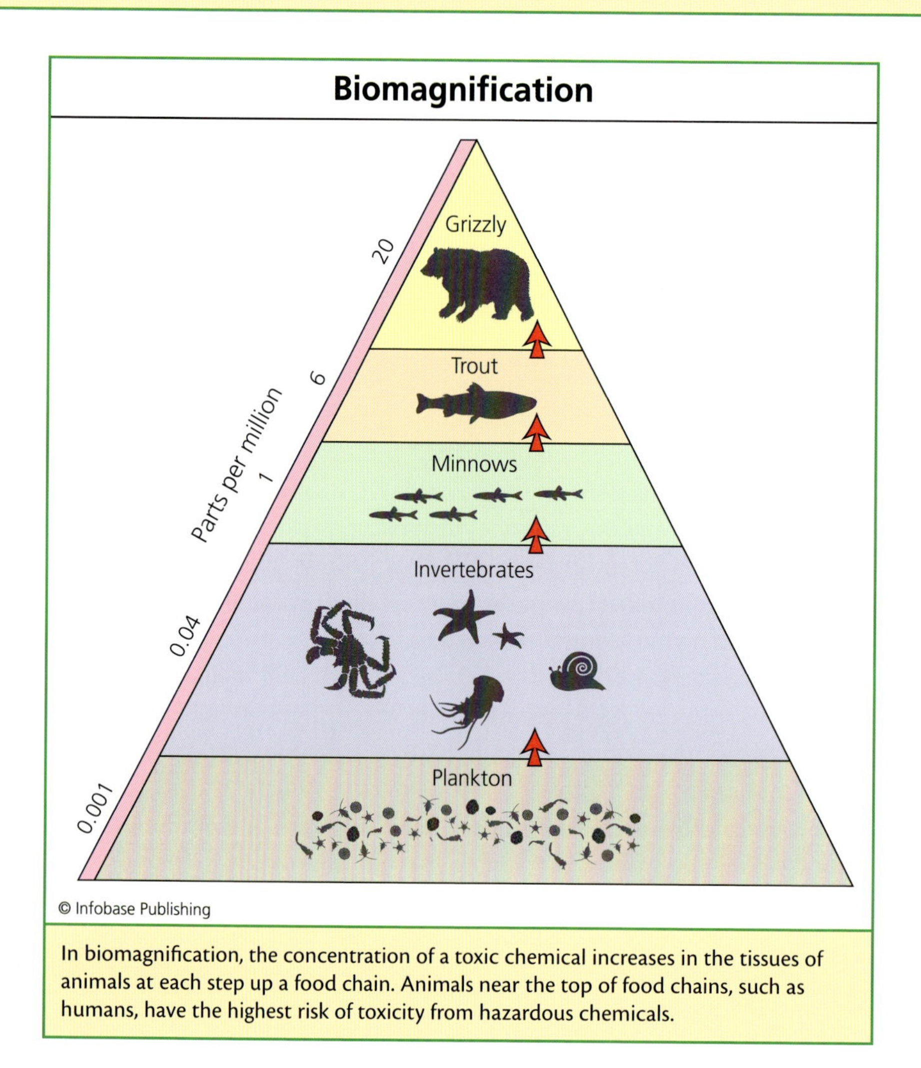

In biomagnification, the concentration of a toxic chemical increases in the tissues of animals at each step up a food chain. Animals near the top of food chains, such as humans, have the highest risk of toxicity from hazardous chemicals.

to determine a person's exposure to a toxin. Second, exposure is usually to more than one toxin, each of which causes different symptoms. Third, the time between exposure to a toxin and the onset of disease can be years or even decades.

The World Health Organization provides scientists and the medical community with information on global diseases caused by environmental toxins. Medical doctors then turn to specialties to treat these diseases and prevent further exposure. Environmental medicine therefore makes use of neurology, detoxification therapy, and surgery among other solutions to environmental disease. The environment offers few easy answers for doctors to use in diagnosing environmental disease. Doctors must call upon a wide variety of sciences in addition to the medical sciences to identify environmental risks.

Environmental medicine also must contend with the problem of biomagnification in which toxins increase in concentration from each food source to the next up the food chain. Due to biomagnification, a toxin's concentration can be very low in single-celled organisms but gradually increase in each consumer and predator until the predator at the top of the chain receives a very large dose of toxin. Humans therefore have a very high risk of illness due to biomagnification.

Environmental medicine will likely grow in importance in the near future, especially if pollution continues. This discipline may depend on advances in epidemiology to make diagnosis easier and faster. Even with advances in medicine and diagnosis, environmental medicine may also depend in large part on intuition, observation, and good fortune in finding and using medical clues.

Epidemiology of Environmental Diseases

The Physician John Snow performed the first epidemiology study documented in medical history in the 1800s by investigating the cause of a waterborne illness in a single area of London. Like scientists who came later in the field, Snow considered the possible toxins, the environment, and illnesses he saw in the local population. In a practical sense, epidemiologists study how a disease enters a population and how it spreads. Like all other aspects of public health, environmental medicine depends on the probability of a disease entering a population.

This chapter covers environmental epidemiology and discusses why finding the source of an environmental toxin may not be as straightforward as physician Snow's detection of a single contaminated water pump. Today the air, soil, and water can be filled with thousands of different chemicals, and human populations move around more than in Snow's time. Epidemiologists therefore have a daunting task in relating a toxin, its source, and possible disease in a mobile society. This chapter discusses the tools of epidemiology that help make these relationships easier to construct. The chapter also provides a basic primer on statistical methods used in epidemiology, new concepts in diagnosis and health, the field of toxicology, and the special challenges of linking environmental toxins with various types of cancers. The chapter also explains the concepts of dose and exposure, bioaccumulation, and persistence of chemicals in the environment and in living things.

EPIDEMIOLOGY

Epidemiology comprises steps that guide investigators from a group of sick individuals to the source of their sickness. Following an illness back to its source is called a *trace-back,* and public health epidemiologists conduct trace-backs in cases of chemical poisonings, water contamination, and food-borne infection outbreaks.

Toxins originate from two types of sources: point sources or area sources. Point sources consist of single identifiable sites, such as a factory, a landfill, a discharge pipe, or a nuclear power plant. Area sources put toxins into the environment from numerous origins, such as vehicle emissions, seepage into aquifers, or waste disposal along a 100-mile stretch of river.

An epidemiology investigation begins by locating, mapping, and interviewing. Investigators locate the area where a disease outbreak has occurred then they map the location of individuals with related symptoms. A group of illnesses confined to a single area suggests that the toxin came from a local point source. A large group of people with similar symptoms, but dispersed across a larger region, points to a toxin from an area source. Finally investigators interview people having the symptoms to narrow down the possible illnesses and disregard other unrelated diseases.

Scientists who work on epidemiological studies of toxins face the chance of exposure to toxins themselves. Many hazardous chemicals are invisible or give off little odor or other clues to indicate their presence in the environment. These firefighters performing a training exercise wear hazmat suits that create a barrier against toxic liquids, solids, and gases.

The following questions help connect an outbreak with a toxin's source:

- Do the people live downwind of a common site, such as a factory?

- Do the people live downstream of a common water source, such as a treatment plant?
- Does the illness concentrate in specific age groups?
- Does the illness concentrate in specific ethnic groups or other subpopulations?
- Do the people who are ill share a common history of jobs, even many years ago?
- Is the incidence of the illness increasing or decreasing?
- Does the incidence increase in a specific area and decrease farther from that area?
- Do the people share certain behaviors: alcohol consumption, smoking, or dietary disorders?

Environmental disease sometimes occurs long after a toxic exposure. For example cancer has a *latency period,* meaning it may not develop into symptoms until years (about 15 to 25), after a person has been exposed to the toxin. Epidemiologists rarely find every person that had been exposed decades ago, so they rely on other evidence for tracing the history of a disease, as follows:

- family members
- vital records—birth and death records
- hospital records
- doctors' reports
- clinical records—blood tests, immunity tests, etc.
- poison control centers
- workers' compensation insurance

Taken together, epidemiology combines science with on-the-ground legwork to find people who had exposure to a toxin, family members, and health records. Even past newspaper articles may report a mysterious outbreak in a community with no known cause. The United States and other countries have created national disease registries, including a registry for cancer incidents and mortalities. The National Cancer Institute's registry provides data on the types of cancers in every state and county. These data help investigators determine if a cancer rate is abnormally high in a

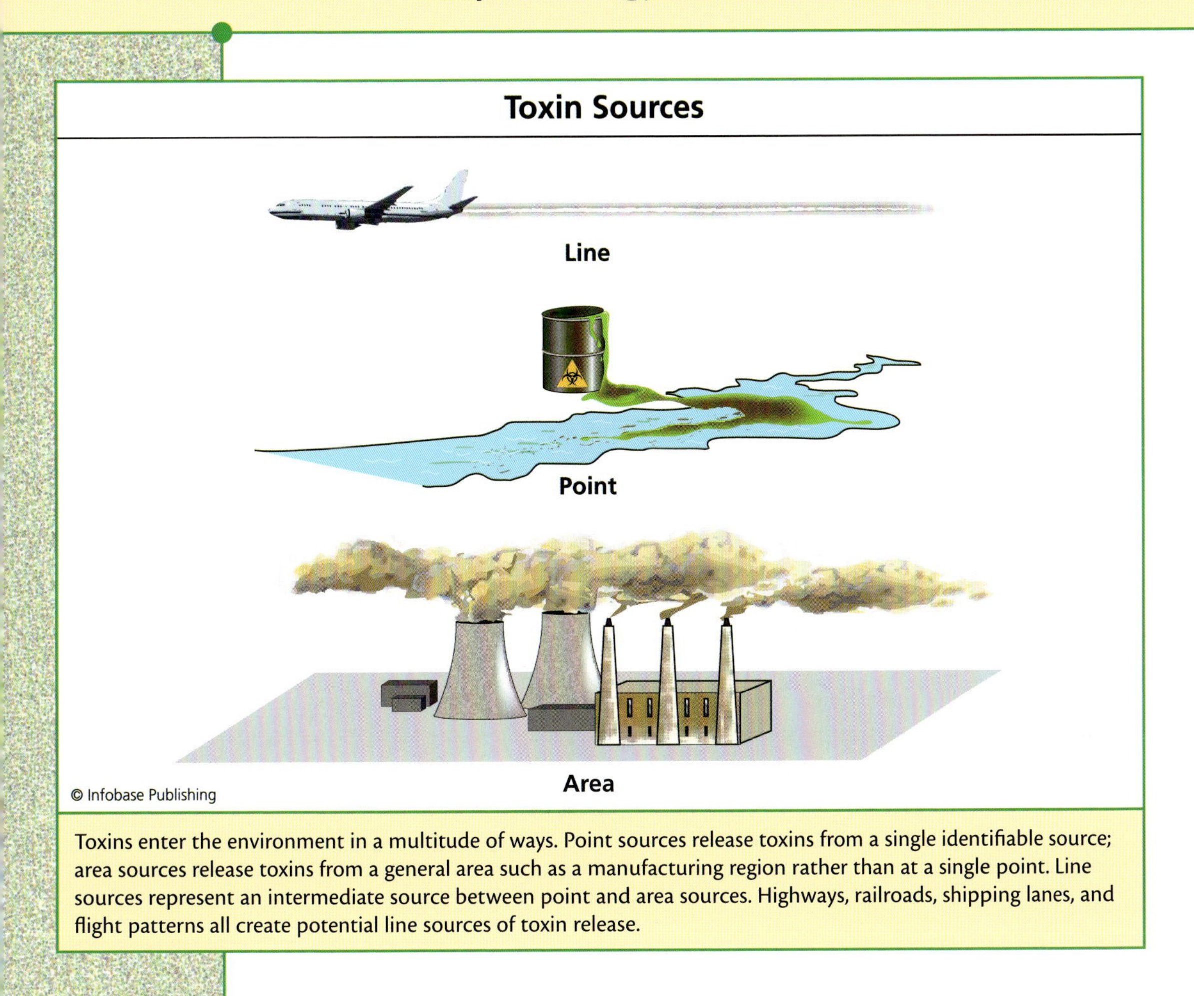

Toxins enter the environment in a multitude of ways. Point sources release toxins from a single identifiable source; area sources release toxins from a general area such as a manufacturing region rather than at a single point. Line sources represent an intermediate source between point and area sources. Highways, railroads, shipping lanes, and flight patterns all create potential line sources of toxin release.

study area. In fact, of the tools used in epidemiology, statistics may be the most important in determining if an illness is truly out of the norm and perhaps connected to toxin exposure.

STATISTICS IN ENVIRONMENTAL SCIENCE

Epidemiology studies use *biostatistics,* data related to the trends and relationships among biological organisms. Biostatistics has also been referred to as *biometry,* a science that measures life. Statistics entered the world of science in the 1600s with the work of two French mathematicians, Blaise Pascal and Pierre de Fermat. Both men developed a system for calculating probability, which is the likelihood that an event will occur. Swiss mathematician Jakob Bernoulli built upon the French studies by expanding on the probability theory to develop the statistics in use today. Bernoulli's posthumous publication *Ars Conjectandi* of 1713 described how probability applied to games of chance such as rolling dice. In 1718 a French student of logic and philosophy living in England, Abraham de Moivre, wrote *The Doctrine of Chance.* De Moivre combined basic statistics with Bernoulli's theory of probability to describe populations. This description became known as a *normal distribution,* often referred to as a bell curve or normal curve. All modern statistics use the foundations laid down by these scientists.

Biostatistics consists of two types of data analysis: *descriptive statistics* and *inferential statistics.* Descriptive statistics describes a population based on its main characteristics. The following provides an example of descriptive statistics to compare two populations:

- The mean ovarian cancer rate in County A is 9 cases per 100,000 people;
- the mean ovarian cancer rate in County B is 13 cases per 100,000 people.

Descriptive statistics therefore enable a scientist to learn about a single population or compare different populations. Descriptive statistics also uses three types of data collections:

- nominal data—data that can be divided into unrelated categories: gender, race, etc.
- ordinal data—data that contains related categories: impoverished communities and malnourished people
- continuous data—variable data, such as pesticide levels in river water, arsenic levels in blood, etc.

Inferential statistics builds on descriptive statistics and then allows a scientist to draw conclusions about two different populations. Inferential statistics also tests two different hypotheses about these populations, as follows:

- The mean ovarian cancer rate in County A is 9 cases per 100,000 people;
- the mean ovarian cancer rate in County B is 13 cases per 100,000 people.

- Hypothesis 1: County B's cancer rate is not significantly different from County A's.
- Hypothesis 2: County B's cancer rate is significantly different from County A's.

The following table (on page 34) explains the fundamentals of descriptive and inferential statistics.

Scientists cannot collect information from every person in a population or a community, so they use sample statistics rather than population statistics. Sample statistics enable scientists to learn about an entire population by studying only a portion of that population. In a presidential race, for example, polls predict the winner based on a sampling of voters. These samples work best when a population follows a normal distribution.

Inferential statistics use two values known as the *p-value* and *significance*. A p-value is the probability of observing a given result. In a horse race, for example, horse A may have odds of 2 to 1 (written 2:1) and horse B has odds of 100:1. The probability of horse A winning is 1 chance out of

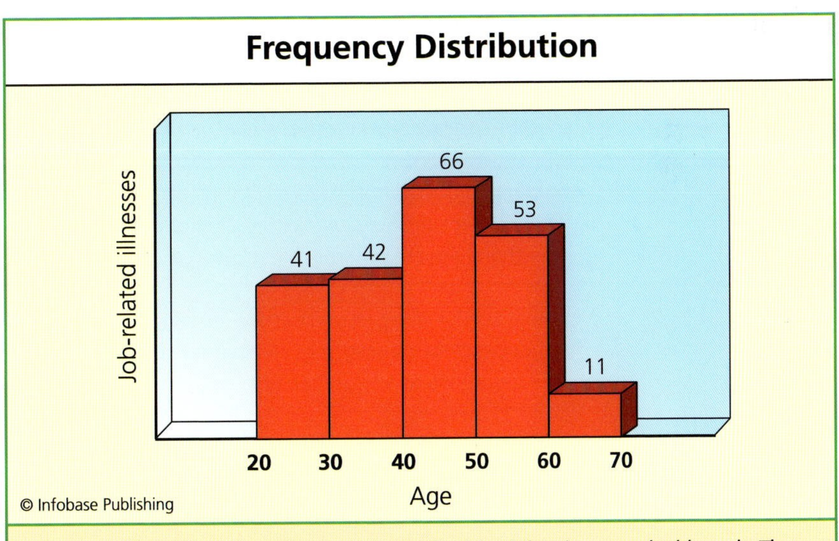

A frequency distribution graph serves as one of epidemiology's most valuable tools. The bar graph distribution shown here indicates the relationship between the age of workers and the probability of getting a job-related illness or injury. Similar distributions have been constructed between specific diseases and many different subpopulation characteristics, including race, income, distance from a toxic site, preexisting illnesses, and so on.

Biostatistics Used in Environmental Epidemiology	
Descriptive Statistics	
mean	the average of a set of values
median	the middle value within a set of values
mode	the most frequently occurring value
frequency distribution	a graph depicting the number of observations recorded at each unit, such as the number of days with a high temperature of 70, 71, 72°F, and so on
range	the difference between the highest and lowest measured values
variance	the measure of dispersion of values around the sample's mean
standard deviation	an indicator of how tightly values cluster around the sample's mean
Inferential Statistics	
t-test	a comparison of two unrelated groups (called independent groups) to determine if their difference is greater than expected or less than expected
paired t-test	a comparison of two groups that are related to each other (dependent groups); e.g., comparing heart rates of people before exercising and after exercising
analysis of variance (ANOVA)	a comparison of more than two groups in which the data are continuous; e.g., the water temperature at discharge pipes of three different nuclear power plants
chi-square test	a comparison of noncontinuous, related data, such as the incidence of lung cancer in cigarette smokers versus nonsmokers

Normal Distribution

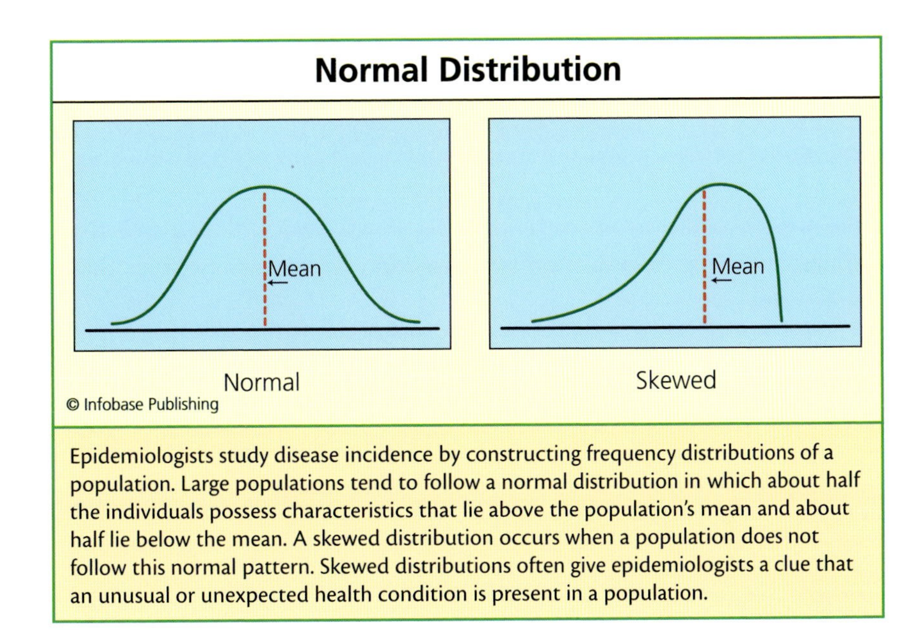

Epidemiologists study disease incidence by constructing frequency distributions of a population. Large populations tend to follow a normal distribution in which about half the individuals possess characteristics that lie above the population's mean and about half lie below the mean. A skewed distribution occurs when a population does not follow this normal pattern. Skewed distributions often give epidemiologists a clue that an unusual or unexpected health condition is present in a population.

2, or 50 percent of the time. Horse B is expected to win only once out of 100 tries; its probability of winning is therefore much smaller (1 percent) than horse A's chances of winning.

Statisticians use the p-value to determine how well a scientific result fits into a normal distribution. For example, a p-value of 5 percent ($p = 0.05$) means a sample has a 95 percent probability of accurately describing the larger population. Likewise, a p-value of 1 percent ($p = 0.01$) assures the investigator that a sample is giving an accurate description of the population 99 times out of 100. The values 0.05 and 0.01 are called *levels of significance,* which help scientists understand the trustworthiness of their sample data. As p-value decreases, the trustworthiness increases. (A p-value of 50—$p = 0.5$—percent means that the sample has only a 50:50 chance of correctly describing the entire population from which the sample came. Considering the horse race, even with a p-value of 0.5, a gambler can depend on horse B losing the race more than he can depend on horse A to win the race.)

Statistics do not give scientists the exact answer to their questions; statistics merely give an idea of how accurately a sampling of a large population describes that entire population. After all the statistics have been calculated, a public health scientist must still use all the information on hand to assess a community's health risks.

ENVIRONMENTAL DISEASES

Different classes of toxins attack living tissue in different manners, and even within these classes, different dangers exist between chemicals. In general, the main groups of environmental toxins listed in the following table target specific organs or tissue types. Environmental medicine similarly groups diseases into categories based on the organs or tissues that a toxin harms.

Prevalent Environmental Diseases

Disease Category	Examples
cancer	any organ or tissue
cardiovascular	heart disease, fatigue, arrhythmia
dermatologic	skin sores, rashes, baldness, dermatitis
endocrine	goiter, hormone disruption
gastrointestinal	diarrhea, vomiting, Crohn's disease
immune	lupus, increased rate of infections, allergies
musculoskeletal	osteoporosis, stunted growth
neurological	autism, Parkinson's disease, Alzheimer's disease, convulsions, headaches, vision loss, memory loss, paralysis, reduced IQ
psychiatric and behavioral	schizophrenia, anxiety, eating disorders, depression
pulmonary	asthma, pneumonia, influenza, emphysema, chronic obstructive pulmonary disease, pneumoconiosis (black lung disease)
reproductive	birth defects, developmental disorders, low birth weights, preterm births, menstrual disorders, ovarian cysts, reduced fertility, impotence
urinary	kidney failure

Advances in detection of the types and amounts of chemicals have improved greatly since the 1970s, as have methods for diagnosing and treating disease. The public must play a role in protecting its safety as well. People should be aware of obvious sites that contain hazards, such as this open-pit uranium mine, but also hidden hazards downwind or downstream from pollution sources, or potentially contaminated groundwater.

Thirty years ago the Surgeon General of the United States, Julius Richmond, warned Congress, "The public health risk associated with toxic chemicals is increasing, and will continue to do so until we are successful in identifying chemicals which are highly toxic and controlling the introduction of these chemicals into our environment." Many chemicals have been identified since then due to advances in analytical equipment, but chemists continue to invent new compounds. All of the potential hazards of new chemicals may take years to understand. Denver toxicologist Scott Phillips cautioned in a 2006 *National Geographic* article, "Any substance, even seemingly harmless ones, can be dangerous in certain quantities and under the right circumstances." With this thought in mind, the public and the medical community must find ways to decrease the chances of exposure to environmental toxins.

The toxins that are easiest to avoid are the toxins people can see. Chemicals pouring from exhaust pipes and smokestacks provide obvious clues about air quality, for example. Global warming has also begun to increase another fairly obvious threat: infectious diseases carried by insects. An infectious disease is any illness caused by bacteria, fungi, or viruses that infect the body. Many of the diseases caused by these pathogens use insects as a mode of transmission, and since global warming enhances insect breeding grounds, insect-borne diseases (called vector-borne disease) may be increasing. The following table lists vector-borne diseases that may increase with global warming.

People can use the information known about environmental diseases to decrease their chances of exposure or infection. The following six actions help decrease the incidence of environmental disease: (1) avoiding visibly polluted areas; (2) understanding local water quality; (3) learn-

VECTOR-BORNE DISEASES

DISEASE	VECTOR	MAIN REGION, CURRENTLY
African trypanosomiasis (sleeping sickness)	tsetse fly	tropical Africa
dengue	mosquito	tropics
dracunculiasis	water flea	tropics
filariasis	mosquito	subtropics
leishmaniasis	sand fly	tropics
Lyme disease	deer tick	North America
malaria	mosquito	tropics and subtropics
onchocerciasis (river blindness)	black fly	Africa/Latin America
plague	rat flea	Africa
schistosomiasis	water snail	subtropics
yellow fever	mosquito	tropical South America and Africa

ing about the origin of certain foods, such as fresh vegetables or imported foods; (4) taking special precautions for people in at-risk health groups; (5) inspecting family history regarding disease; and (6) becoming aware of a local area's past, such as the presence of industry and the type of industry.

At-risk groups are especially vulnerable environmental illnesses. People in high-risk health condition, meaning bodily systems are not at their optimal, have increased risk factors to the onset of disease. Risk factors are among several components in epidemiology that influence the *survivorship curve,* described in the sidebar below.

The Survivorship Curve

A survivorship curve is a graph that depicts the number of individuals expected to survive or die due to a single factor. Survivorship curves most often portray the effects of age on the survivability of humans or other animal species. Environmental medicine often uses survivorship curves to illustrate the effects of a toxin on mortalities of different age groups. These graphs may be used in the two following ways: (1) to pinpoint the most vulnerable times in a person's life after exposure to a toxin, and (2) to compare the mortality rates of two populations, one of which was exposed to a toxin and the other population not exposed.

Three types of survivorship curves portray the normal life span of all individuals in a population. A convex curve (type 1 survivorship curve) represents biota that have a high survival rate among the young and individuals live to an old age, such as humans or elephants. A fairly straight line or constant survivorship curve (type 2) describes biota with a constant death rate throughout the life span of the species. Coral, reptiles, honeybees, and squirrels follow a constant curve. A concave curve (type 3) represents biota with high death rates for the young, but the survivors live to an old age. Plants, oysters, sea urchins, and loggerhead sea turtles provide examples of concave survivorship curves.

A survivorship curve cannot describe all the ways in which a toxin affects life span and mortality rates, but this tool can help epidemiologists assess how toxins harm entire populations.

The Survivorship Curve

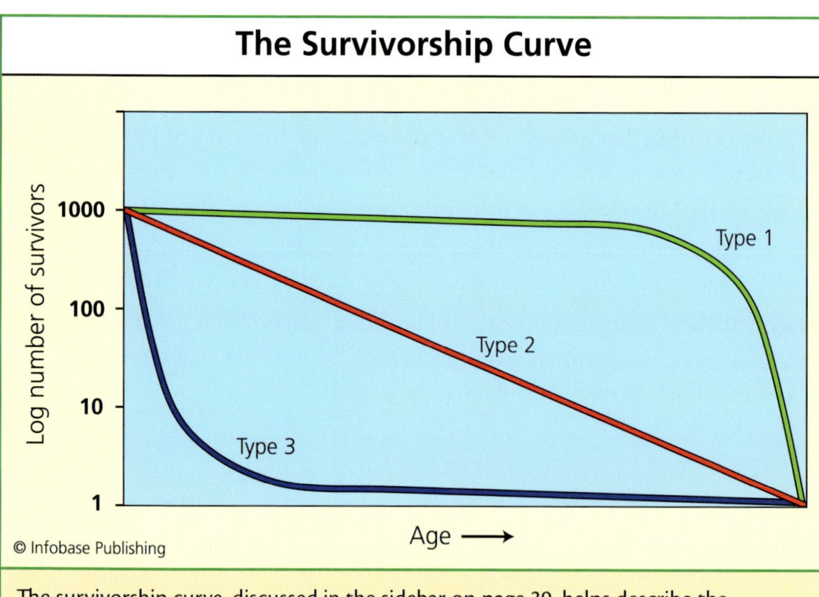

The survivorship curve, discussed in the sidebar on page 39, helps describe the life expectancy of various animals, including humans. Epidemiologists can also create survivorship curves of subpopulations in the human population. The world's population follows a type 1 (or I) pattern, but regions of the world with very high child mortality rates may follow a pattern described more by type 3 (or III) survivorship than type 1. This gives medical care providers important information on the vulnerability of a subpopulation to disease.

DIAGNOSIS

Physicians diagnose environmental diseases in the same way they diagnose any other type of disease, in three steps: (1) listening to the patient describe the symptoms; (2) performing an examination and laboratory tests to assess the patient's physical condition; and (3) correlating the symptoms, findings of the exam, and test results to the characteristics of known diseases. These steps proceed with few problems unless the disease has never been seen before in medicine. A 1976 event in Philadelphia, demonstrated the problems that arise when an unknown environmental toxin enters a population. The event involved an American Legion convention in which members became sick from an unidentified agent; two died. Doctors followed standard diagnosis procedures, but their findings did not match the symptoms of any known diseases. Investigators took six months to find the causative agent, a bacterium that would be named *Legionella,* and determined that the convention hall's cooling system acted as the bacteria's point source. (This organism had already been discovered, but microbiologists associated the microbe only with animals and not humans.) Pennsylvania secretary of health Leonard Bachman explained the situation to the *New York Times* in 1976: "We thought we might be faced with an unprecedented condition in modern medicine, one for which we had no antibiotics, drugs or therapy." These fears arise any time a new disease enters a population.

The period between the time a sick patient sees a doctor and the doctor's diagnosis can be very tense, as it was in the Legionnaire's disease outbreak. In environmental medicine, patients may furthermore have been exposed to several toxic chemicals rather than a single toxin, making diagnosis all the more difficult. A varied mixture of toxins that give rise to a complex set of symptoms is known as *multiple chemical sensitivity* (MCS).

Experts in environmental medicine have suggested that MCS may relate to genetic makeup, causing certain individuals to have a high sensitivity to toxins. Some physicians feel, however, that MCS's collection of undefined symptoms do not correlate with any known disease. Ronald E. Gots, a physician who specializes in MCS investigations, has stated that MCS "is a label given to people who do not feel well for a variety of reasons and who share the common belief that chemical sensitivities are to blame. It [MCS] defies classification as a disease. It has no consistent characteristics, no uniform cause, no objective or measurable features. It exists because a patient believes it does and a doctor validates that belief." The debate about the true nature of MCS continues in environmental medicine.

In 2007 the online medical resource ImmuneSupport.com interviewed biochemistry professor Martin L. Pall of Washington State University, a proponent of the existence of MCS as a real medical condition. He said, "There are stressors that are most commonly involved in the initiation of one illness but rarely involved in the initiation of others. For example the organic solvents and the pyrethroid and organochlorine pesticides seem to be most commonly involved in the initiation of Multiple Chemical Sensitivity." Edward B. Holmes, director of the Environmental Health Clinic at the University of Utah, summed up the topic for Colorado's *Fort Collins Journal*, "The bottom line is that the condition is very much in dispute." Illnesses such as MCS that present a variety of vague symptoms challenge all fields of medicine. Some health professionals have recently developed a new approach to managing hard-to-define diseases caused by unidentified toxins. This approach is highlighted in the following sidebar, "Case Study: New Views on Homeostasis."

TOOLS USED IN DIAGNOSING DISEASE

Medical care providers diagnose disease in patients using a mixture of soft skills (listening, asking questions, and observing behavior) and hard skills (examining the patient, running tests, and using medical knowledge and experience). An accurate diagnosis then results from one or more of the following activities: surveying the patient's family history, physically examining of the patient, and testing.

Medical histories provide information on the genetic factors that could make a person susceptible to disease such as heart disease or cancer. A patient's medical history provides doctors with clues about a diseased state, but likely cannot give enough information to make a diagnosis. Physical examinations help the doctor assess a patient's overall health. An examina-

CASE STUDY: NEW VIEWS ON HOMEOSTASIS

The American Academy of Environmental Medicine (AAEM) was founded in the 1960s to gather information on numerous symptoms and illnesses that had been emerging in communities, possibly because of environmental toxins. The AAEM took the unique approach of blending the strengths of Western medicine with those of Eastern medicine—both disciplines refer to themselves as "traditional" medicine. As a result, the AAEM proposed a novel way to define *homeostasis*.

In Western medical schools, homeostasis refers to the actions the body takes to adjust to the external environment. The body constantly adjusts to heat, cold, humidity, light, darkness, and noise, to name a few stimuli. In circumstances when the body cannot adjust to the environment, injury or illness may occur. Western medicine does an excellent job treating acute illnesses and trauma, but critics point out that Western medicine puts less emphasis on the body's small and constant responses to stress, which may affect how the body fights disease. The AAEM describes this approach to homeostasis in the following way: "The environment is seen as an essentially benign place that generally has little effect on health, and the diet is simply a passive source of metabolic fuels for [the body]. . . ." This description intimates that the AAEM holds a different view of how to manage the body's response to environment.

tion begins with measurement of a patient's height, weight, and temperature. Temperature indicates if the body is maintaining homeostasis by a process called thermoregulation, in which the body monitors and adjusts its internal temperature. Elevated temperature indicates the possible presence of an infection. Physical examinations fall into the following categories:

- respiratory—respiratory rate, lungs, trachea, sinuses, ease of breathing
- cardiovascular—heart rate, pulse, blood pressure, sensation in extremities
- abdominal—tenderness, pain, abnormal internal growths
- enteric—bleeding, diarrhea, constipation, pain
- neurological—reflexes, sensation, hearing and smell, balance, headaches, memory
- urinary—bleeding, signs of infection, pain in urinating

Blending Western with Eastern philosophies creates a new view of homeostasis. The body reacts to external factors such as heat and cold, but it also reacts to stressors that enter the body. Stressors such as nutrients strengthen the body's functions while stressors such as environmental toxins strain the body's normal functioning. Each person furthermore possesses his or her unique sensitivities to all the different stressors faced in a lifetime. This expanded approach to homeostasis emphasizes diet and a healthy environment as it emphasizes disease treatment. In other words, disease prevention becomes as important as disease treatment.

Medicine remains an evolving science, which has not yet defined the correct actions to take for all of the world's illnesses. Therefore, a shift in thinking about medical care will probably not come easily. Environmental health specialist Edward B. Holmes reveals the barriers that may still exist in looking at disease in a new manner: "There seems to be, in my experience, a significant number of people that have this kind of conglomeration of symptoms [from environmental toxins] that fit into a pattern strongly with psychiatric conditions." Put another way, some doctors may continue to believe MCS is "all in your head." The true nature of MCS and even its existence appears to remain a question for the future, but before answering these questions about MCS, people can promote better health by eating a healthy diet and avoiding known toxins.

- genital—pain, signs of infection
- dermatological—rashes, redness, nodules, growths, discoloration, hair loss
- ophthalmic—sight, blurred vision, loss of peripheral vision
- lymphatic—swollen glands, painful glands, breast exam

A physical examination includes also a test of abnormalities in the blood or other specimens, which requires specialized equipment to analyze blood constituents. The table below lists the major constituents measured in a complete blood analysis used in environmental medicine. The medical meanings of these measurements are explained in appendix A.

CONSTITUENTS OF A COMPLETE BLOOD COUNT (CBC) AND BLOOD CHEMISTRY TESTS

CONSTITUENT	MEASUREMENTS
cells	red blood cells (RBC), hematocrit (Hct), platelets, mean corpuscular volume (MCV), red cell distribution width (RDW), white blood cells (WBC), neutrophils, lymphocytes, monocytes, eosinophils, basophils
antigens and other disease markers	blood type antigens, tumor/cancer markers, specific proteins or enzymes
elements	sodium, potassium, chloride, calcium, normalized calcium, blood urea nitrogen (BUN)
compounds	hemoglobin (Hgb), mean corpuscular hemoglobin (MCH), mean corpuscular hemoglobin concentration (MCHC), total protein, albumin, globulin, carbon dioxide, glucose, bilirubin, creatinine
enzymes	alkaline phosphatase, aspartate aminotransferase, alanine transaminase
lipids	cholesterol, triglyceride, high-density lipoproteins, low-density lipoproteins
other	glomerular filtration rate (GFR), anion gap

In addition to blood samples, medical examinations include testing of additional specimens—feces (stool), sputum, urine, skin, hair, and cerebrospinal fluid. Finally, medical diagnoses rely on instruments to analyze the constituents in specimens and to allow doctors to "see" inside the body to look for abnormalities in organ size and density or to spot unusual growths such as tumors. Appendix B lists the common instruments used in Western medicine for diagnosing disease.

TOXICOLOGY

Toxicology is the study of the harmful effects of chemicals on health. The field of toxicology also contains specialty areas including renal (kidney) toxicology, reproductive toxicology, neurotoxicology, and hepatic (liver) toxicology. Two important components of toxicology relate to the way in which an environmental toxin affects a living organism: the substance's toxicity and its *dose.* Toxicity is the measure of how harmful a substance is to living tissue. Dose contributes to a substance's toxicity; dose is the measured amount of a toxin put into a living body. Five other factors that affect toxicity are: (1) the length of time the body has been exposed to the toxin; (2) the number of times the body has been repeatedly exposed to the toxin; (3) the risk status of the individual; (4) the genetic makeup of the individual; and (5) the current ability of the body to repel injury. A person will repel damage from toxins better if the kidneys, liver, lungs, heart, or immune system are working at their best at the time of exposure to the toxin.

Another important factor in determining the toxicity of a substance arises from the physical characteristics of toxins. These characteristics determine how readily a toxin will enter a body and cause harm once it is on or in the body. The following five factors influence the harmfulness of environmental toxins:

1. Water-solubility versus fat-solubility: Water-soluble toxins spread in the environment in rain, drinking water, surface waters, and groundwaters, and they also spread through the body—water makes up 55 percent (in females) to 65 percent (in males) of the human body. Fat-soluble toxins, by contrast, stay in oily substances in the environment and penetrate cell membranes in the body and persist in the body by residing in fatty tissue.

2. Persistence: Persistent toxins stay in the environment a long time—up to several decades—and so increase the chances of exposure and length of exposure time.
3. Bioaccumulation: Toxins that persist in living tissue have a greater chance to accumulate to higher and higher amounts in the body over time.
4. Biomagnification: Persistent toxins that accumulate in tissue also tend to increase in concentration up a food chain. Predators at the top levels of food chains therefore are at greater risk of harm than prey at the lower levels of the food chain.
5. Interactions between toxins: Some toxins become more harmful in the presence of another substance. Cigarette smokers, for example, run a greater risk of lung disease from environmental toxins in the air.

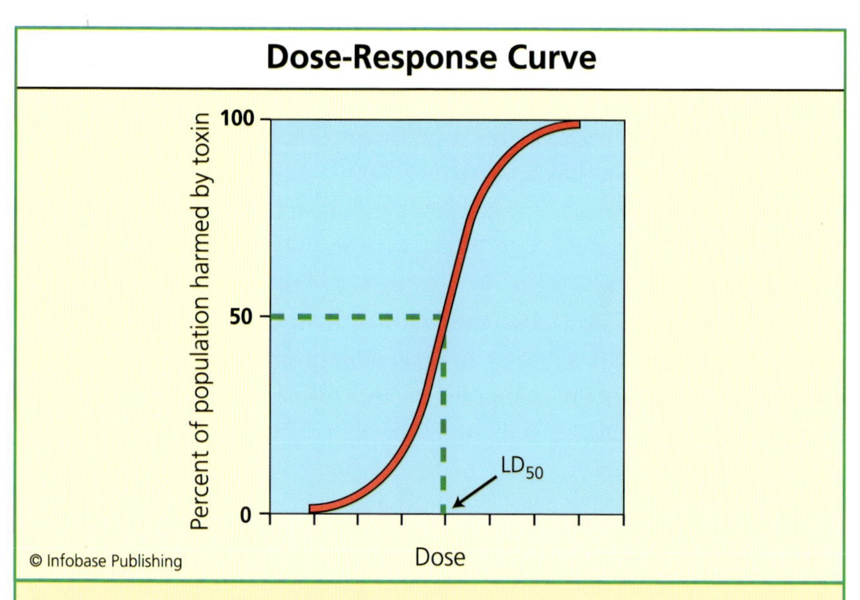

The dose-response curve enables scientists to determine the LD_{50} of any toxin. This dose-response calculation works as shown here when a population has a normal or near normal distribution, and enables the scientists to make direct comparisons between the toxicities of different chemicals. The lower the LD_{50}, the lower the dose needed to cause illness in a normal population. In other words, a low LD_{50} indicates a chemical is very toxic.

The measurement of toxicity is called the *median lethal dose,* which is the dose of toxin that kills 50 percent of test animals (usually mice) within 14 days in a controlled dosing experiment. This calculation uses 50 percent as a benchmark rather than 100 percent for the purpose of eliminating any variability between organisms. Even the most powerful poison unleashed on an entire community would be unlikely to kill every member of the community. Achieving 100 percent mortality also does not give much information about the lethal dose; perhaps the dose that is shown to cause 100 percent of animals to die is 10 times greater than needed. The median lethal dose, abbreviated or LD_{50}, therefore provides information whereby toxicities can be compared, usually in milligrams per kilogram (mg/kg). (It may also be written as LC_{50} for "lethal concentration.")

Toxins vary greatly in their LD_{50} values. Nerve gases usually have an LD_{50} of less than 1/100,000,000th of a pound of body weight (0.01 mg/kg); the organic solvent carbon tetrachloride, by contrast, has an LD_{50} of up to 5/1,000th of a pound of body weight (5,000 mg/kg). Put another way, the lower the LD_{50}, the more toxic the substance. Substances with low LD_{50} values require only a small dose to be toxic. Any substance can cause death at a high enough dose, even water.

CHRONIC EXPOSURE

People do not always receive just a single dose of an environmental toxin, like the dose used to calculate LD_{50}. Exposure to toxins more often occurs over many weeks or years and the exposure can be either continuous or repeated. For example, a family living next to a toxic waste dump might receive constant exposure to the site's chemicals. By contrast, a contractor who demolishes buildings may be exposed to asbestos fibers only a few times a year.

Two types of toxicity arise in biota based on these kinds of exposure: (1) *acute toxicity* from a single large dose of toxin, or (2) *chronic toxicity* from a long exposure period or repeated exposures to small doses. Chronic toxicity may also occur from a single dose of a toxin that stays in the body and causes harm for a long time afterward.

Chronic exposure to toxins is the same as chronic sublethal poisoning, discussed in the previous chapter. In any kind of exposure, the amount of chemical to which a person is exposed has a great impact on health. University of Kansas toxicologist Karl Rozman told *National Geographic* in 2006, "In toxicology, dose is everything." Industrialized countries today have the potential of exposing residents to a grab bag of chemicals. Some of the most common chemicals found in people's bodies in the United States are the following: flame retardants such as polychlorinated biphenyls (PCBs) and polybrominated diphenyl ethers (PBDEs); phthalates used in making plastic; bisphenol A used as a coating in water bottles and food containers; formaldehyde that vaporizes from furniture finishes; the degreasing solvent trichloroethylene (TCE); and tetrachloroethylene (also known as perchloroethylene, PERC, or PCE) used in dry-cleaning.

Two additional sources of chronic exposure have been a concern since at least the 1960s: air and water. Air pollution has improved in some regions of the United States but remains the source of dangerous substances here and abroad. Surface waters carry many industrial chemicals and pesticides, even pesticides that have been banned but remain in the environment. A U.S. Geological Survey (USGS) study that ran from 1992 to 2001 found the pesticides dichlorodiphenyltrichloroethane (DDT), dieldrin, and chlordane in streams even though these pesticides have been banned for several years. The USGS's Robert Gilliom said, "The potential effects of contaminant mixtures on people, aquatic life, and fish-eating wildlife are still poorly understood and most toxicity information, as well as water-quality benchmarks used in this study, has been developed for individual chemicals. . . . studies of mixtures [of contaminants] should be a high priority." Gilliom is correct in stating that toxicology must focus on the effects of chronic exposure to chemical mixtures in the same way it studies single chemical exposures.

Chronic exposure to a chemical lasting for a year or longer presents special health concerns. A chemical that is not particularly dangerous may become more dangerous as people are constantly exposed to it. Migrant workers, such as the man shown here spraying pesticides with his family standing nearby, have a potentially higher likelihood of chronic exposure to chemicals.

Bioaccumulation and Persistence

Bioaccumulation is the gradual increase of a chemical in the tissues of a living organism. Chronic exposure to environmental toxins may lead to their buildup in certain organs if tissues cannot excrete the toxin and the body cannot eliminate it in feces, urine, or sweat. This process of the body ridding itself of a toxin is called *clearance.*

A chemical that the body cannot clear or degrade fast enough is called a *persistent chemical,* which then accumulates in the body. Four factors contribute to the persistence of chemicals in the bodies of humans and other animal life: (1) animals with large amounts of fat, which retains fat-soluble chemicals; (2) long-lived animals that retain chemicals for many years; (3) animals with slow metabolism that retain chemicals longer than animals with fast metabolism; and (4) chronic exposure.

On May 23, 2001, the United States joined more than 100 other countries in signing the Convention on Persistent Organic Pollutants (POPs). This treaty focused on setting goals for reducing or eliminating the use of POPs. The treaty further listed 12 persistent substances, nicknamed the "dirty dozen," that were of greatest concern if people received chronic exposure to them. These POPs are listed in the following table.

POPs of Greatest Concern in Environmental Health
Pesticides
aldrin, chlordane, DDT, dieldrin, endrin, heptachlor, mirex, toxaphene
Industrial Chemicals
hexachlorobenzene (HCB), PCBs
By-products of Industrial Processes
polychlorinated dibenzo-p-dioxins (dioxins), polychlorinated dibenzo-p-furans (furans)

(continues)

(continued)

Persistent chemicals like the dirty dozen are usually large and complex compounds that microbes in the environment cannot break down. As scientists learn more about the hazards of synthetic chemicals and as chemists learn to better identify them in soil and water, the list will probably expand. But knowledge is a powerful aid in protecting health, so the more scientists learn about POPs, the better chance they have in detecting these chemicals and improving environmental health.

Chronic exposure to toxin mixtures by young children presents an added concern in environmental health. Medical researchers believe that exposures that appear to cause no harm to babies may lead to health problems later in life. A team of scientists was quoted in a 2007 *San Francisco Chronicle* article on chronic exposure: "Given the ubiquitous exposure to many environmental toxicants, there needs to be renewed efforts to prevent harm. Such prevention should not await detailed evidence on individual hazards." Team member and pediatrician Philip Landrigan added, "A sad aspect with many of these prenatal exposures is that they leave the mother unscathed while causing injury to her fetus." Clearly chronic exposure presents a health threat to certain populations and age groups living near large amounts of pollution. On the other hand, the animal body has the capacity to eliminate toxins over time to preserve overall health. Chronic exposure therefore contains many issues that have not yet been completely studied. The "Bioaccumulation and Persistence" sidebar on page 50 describes one important facet of chronic toxin exposure.

POLLUTION AND CANCER

Toxic chemicals can be categorized as mutagens, teratogens, or carcinogens, depending on their effect on the body's physiology. Mutagenic substances cause mutations in living things by damaging deoxyribonucleic acid (DNA). Teratogenic toxins damage developing fetuses, causing birth defects. Carcinogenic toxins cause the growth of malignant tumors and cancers. In the United States, the American Cancer Society has predicted that males of all ages have a 1 in 2 chance of getting *invasive cancer* in their lifetime; females have a 1 in 3 chance, although doctors do not yet know the extent to which environmental factors cause these cases.

The link between cancer and environmental toxins is known in some cases, suspected in others, and in some instances no link at all has been found between a chemical and cancer. People should therefore not assume that every chemical in the environment poses a serious heath threat. National Cancer Institute epidemiologist Aaron Blair said in a 2004 interview, ". . . many substances that we suspected would cause cancer in animals actually do not. Of course, it is possible that they do cause cancer in humans, but, in fact, our experience has shown us that most of the chemicals we have tested don't cause cancer." Massachusetts Institute of Technology chemist Gerald Wogan supported Blair's opinion in a 2005 *New York Times* article: "People differ greatly in their response to chemical carcinogens. Almost all chemicals, with relatively few exceptions, have to be converted from what they are into something more chemically active to be carcinogenic. If you encounter one of these compounds, most of it is converted to less toxic material that is excreted. Only a tiny amount is converted to a form that can cause cancer. A small fraction of 1 percent gets converted." These scientists offer reassurances that the public can temper their fears about environmental toxins until science learns more about these chemicals.

Not everyone has been convinced that science can solve the ongoing dispute about whether the environment is a dangerous place. Barbara Brenner, executive director of Breast Cancer Action, reminded the *Times* that new chemicals enter the environment every day. "Nobody can keep up," she said, "and we don't know the health effects. I think it is not an irrational response to say our environment is making us sick." Brenner may be correct in worrying that not every chemical has been tested; studies on chemical mixtures lags even further behind. Despite these real concerns, the connections between cancer and environmental toxins remain only partially understood. Science rather than emotions will clarify the ways in which environmental chemicals affect people. In the meanwhile, people can use good common sense to avoid obvious hazards and learn as much as possible about the environment. These may be the best ways to protect health.

CONCLUSION

Environmental medicine has made enormous strides in connecting the possible causes of disease with factors in the environment. Most of these connections have been drawn by using the specialized science contained in epidemiology. It is difficult to build an exact cause-and-effect picture to relate a given chemical to a defined illness. Environmental medicine therefore uses the basics of epidemiology—observations of large populations, statistical methods, and laboratory experiments—to build a body of evidence regarding a chemical and illness.

Environmental toxins have alarmed the public for more than four decades as scientists uncover more and more places that hold contamination. In recent years another upsetting discovery has contributed to people's worry: environmental chemicals can be found in many people's bodies. In some instances doctors can predict a disease based on the presence of a chemical in the body, but in many more instances medicine has simply not accumulated enough data to say what chemicals and what doses are required to cause disease. Scientists furthermore do not always agree on the best ways to link chemicals to disease. It is little wonder that the public can be confused and even alarmed about the hidden health threats in the environment.

The study of medicine is part of biology, and biology represents a science that contains few absolutes: biological things vary from individual to individual. People's bodies react differently to toxin exposure for various reasons. First are the genetic differences that make some people more susceptible to illness than other people are. Second, high-risk health groups have a greater chance of being susceptible to toxins than the normal population. Third, toxins do not always behave exactly the same way in different people. Finally, different toxins accumulate in the body to differing degrees, so their consequences may be short-term or chronic, and these consequences also differ.

Medical epidemiology seeks to address all of these variables by using probabilities rather than discrete numbers. Over time, doctors learn about disease by studying the epidemiology of a disease within large populations. The final diagnosis of disease arises from a collection of clues from symptoms, test results, family histories, and physical examinations that a doctor uses to identify the disease. Medicine may be considered an evolving field because the effects of disease vary from person to person and because science constantly expands with new information about the environment, yet it depends in large part on intuition and observation.

Environmental medicine has a responsibility to inform the public of potential hazards, but it must also teach people about the likelihood of exposure to these hazards. The public at the same time would do well to learn as much as they can about potential threats from environmental toxins, their health effects in the body, and the ways of avoiding many of these substances. With good information in hand, people will realize that they may not need big changes in their activities to reduce their exposure to potential dangers. In at least one regard, scientists and nonscientists are similar: Both make their best decisions by relying on common sense.

Environmental Toxins

People and wildlife receive environmental toxins from the air, water, and from the land. In general, toxic substances in air tend to be metal-containing particles and gases; toxins in water tend to be soluble organic compounds and biological substances; and toxins on land are metals, organic chemicals, pesticides, and biological substances. Toxins of biological origin in the environment consist of viruses, bacteria and their toxins, algae and their toxins, protozoa, cysts, fungi, and parasites. These entities come from animal wastes, sewage, or improperly treated wastewaters and may be more of a health threat in developing parts of the world than in developed countries. The World Health Organization (WHO) stated in its *World Health Report 2007,* "Infectious diseases are not only spreading faster, they appear to be emerging more quickly than ever before. Since the 1970s, newly *emerging diseases* have been identified at the unprecedented rate of one or more per year. There are now nearly 40 diseases that were unknown a generation ago. In addition, during the last five years, WHO has verified more than 1,100 epidemic events worldwide." Many of the emerging diseases referred to by the WHO are, however, communicable diseases rather than environmental. Whether chemical or biological in origin, environmental toxins spread across the world either by natural conveyances or with people. Global travel and commerce take place on a large scale. Airplanes and ships that travel halfway around the world can take hazardous substances, especially biological substances, with them.

Biological toxins cause infections in people after they enter the body. Most often the infections lead to fever and digestive upsets. Chemical toxins exert a different type of damage in any of the body's metabolic systems, but they tend to do the most damage to the nervous, immune, and endocrine systems. Severe chemical toxicities of the nerves and brain cause memory loss, loss of coordination, and large doses can lead to paralysis and death. Radioactive chemicals tend to destroy the immune system, leaving a person open to infection from biological agents. The network of glands that comprise the endocrine system secrete hormones that help the body's organs function properly, but a number of industrial chemicals interfere with hormone activity. The topic of *endocrine disrupters* is relatively new in environmental medicine, but evidence has been mounting on the dangers of these compounds to both wildlife and humans.

The WHO has pointed out the importance of globalization in spreading any type of environmental hazard. Globalization occurs when industries own operations and do business in many different countries. "As well as the international mobility of people," the WHO has reported, "the global movement of products can have serious health consequences. The potentially deadly risks of the international movement and disposal of hazardous wastes as an element of global trade were vividly illustrated in Cotê d'Ivoire [Ivory Coast] in August 2006. Over 500 tons of chemical waste were unloaded from a cargo ship and illegally dumped by trucks in different sites in and around Abidjan [Cotê d'Ivoire's largest city]. One month after the dumping, almost 85,000 consultations had been recorded at different medical facilities in relation to the chemical incident and its consequences: 69 people had been admitted to [the] hospital and eight deaths had been attributed to the event." Incidents like this one point out the threats that come from moving large amounts of materials around the world.

In 2005 environmental health researchers Donald Wigle of the University of Ottawa in Canada and Bruce Lanphear of Cincinnati's Children's Hospital Medical Center told *ScienceDaily*, "The public depends on decision makers, scientists, and [government] regulators to restrict exposure to widespread toxins that have known or suspected serious potential health effects." Government leaders also need accurate information on environmental toxins. This is a difficult task because environmental science reveals new information on chemical and biological toxins on an almost daily basis.

This chapter explores the processes by which environmental toxins enter and harm the body. The chapter reviews common toxin groups consisting of pesticides, organic solvents, and metals. It also covers the current knowledge in toxin activity at the tissue and cellular level. Finally,

Environmental toxins have always been present throughout human history, but their variety and concentration have increased since the industrial revolution. Today's chemical manufacturers, power plants, and transportation release more wastes into the environment than ever before. This chemical barge, plying a river, provides an example. The barge is a potential source of a chemical accident, fuel leaks, paints and coatings, and air pollution that could enter the river ecosystem. *(U.S. Fish and Wildlife Service)*

chapter 3 offers information on detoxification methods, therapies, and chemical toxicity prevention.

TOXIN ACCUMULATION ON EARTH

New chemicals enter the environment every day. Almost all of these chemicals have been synthesized in laboratories, and enzyme systems in nature rarely degrade them. When large, complex molecules do degrade in nature, they often break apart into unique structures that persist in the environment. Chemists classify environmental chemicals based on the chemical's original structure. By grouping chemicals this way, scientists study the details of a few chemicals from each group and assume that all chemicals in a group behave similarly when they are in the body. For example, by studying the effect of pesticide A1 on the nervous system, scientists also learn about pesticides A2, A3, and so forth. This approach to chemical testing therefore gives an estimate of many chemical toxicities rather than an exact answer.

The U.S. Environmental Protection Agency (EPA) monitors the categories and many individual chemicals that enter the environment each year. The EPA then makes this information public through its *Toxics Release Inventory* (TRI) Program. The TRI list contains more than 650 chemicals. Any emitter—manufacturing plant, small business, mine, etc.—must tell the EPA about any chemical it has released into the environment and the amount. The Centers for Disease Control and Prevention (CDC) also list many of the same chemicals (and chemical groups) that appear on the TRI, summarized in appendix C.

Analytical chemistry has developed sensitive equipment to detect hundreds of chemicals released into the environment, but environmental monitoring nevertheless presents the following hurdles:

- Industrial chemicals exist in varying amounts in the environment.
- Many chemicals occur in very tiny quantities, called *trace amounts.*
- Chemicals have varying persistence, so that some degrade faster than others.
- Many chemicals undergo partial breakdown and the intermediate products are difficult to detect.
- Some emitters probably do not notify the EPA of releases.

More than 20 years ago, *Time* magazine reporter Ed Magnuson wrote, "At last count, nearly 50,000 chemicals were on the market . . . almost 35,000 of those used in the United States are classified by the federal EPA as being either definitely or potentially hazardous to human health." Today, industries produce about 80,000 different chemicals of varying danger to humans and wildlife. Jay Vroom, president of the American Crop Protection Association and a champion for pesticide use, remarked in a Public Broadcasting Service (PBS) interview, "I think that what we know so far indicates that there is no cause for concern." In the same interview, toxicologist Stephen Safe of Texas A & M University added, "It is incumbent on the scientists, media, legislators, and regulators to distinguish between scientific evidence and hypothesis, and not to allow a 'paparazzi science' approach to these problems." The views expressed by Magnuson and Safe make it clear that science does not hold all the answers it needs on chemicals and health. Scientific debate continues on the following three main aspects of environmental chemicals: (1) the amount in the environment; (2) the amount in human and animal bodies; and (3) the relationship between the presence of a toxin in the body and illness from that toxin.

Scientists use biomonitoring to find answers to these questions and also to study accumulation in people from different age groups. Biomonitoring usually involves the following samples: blood, serum, saliva, urine, breast milk, expelled air, hair, nails, fat, bone, and other tissues such as skin.

In 2006 reporter David Ewing Duncan participated in a biomonitoring study for a *National Geographic* article. The analysis of Duncan's blood and urine showed that he—having had no contact with industry

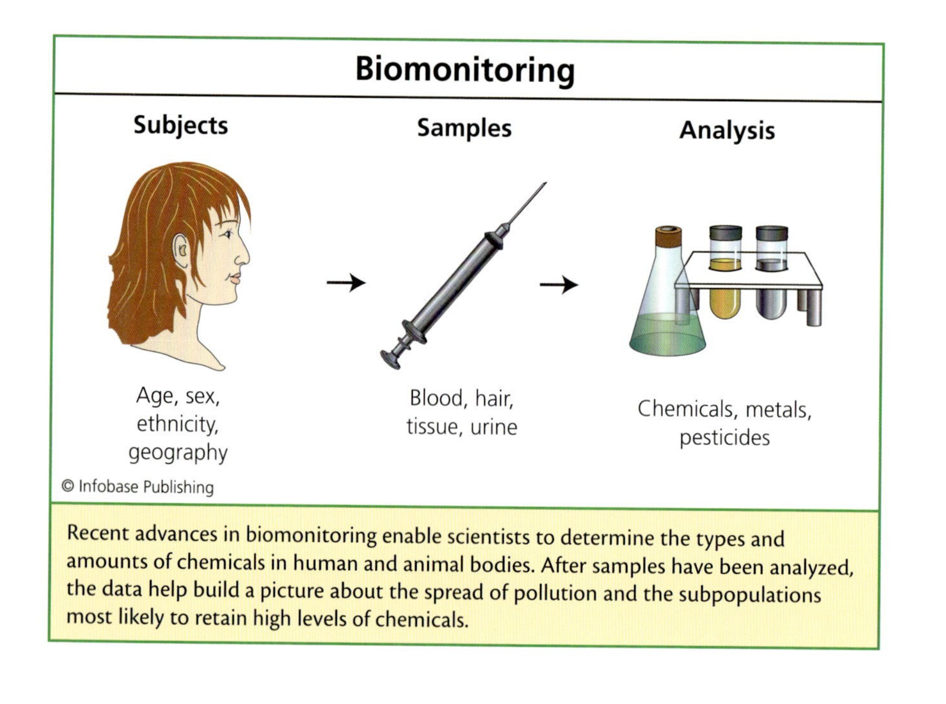

Recent advances in biomonitoring enable scientists to determine the types and amounts of chemicals in human and animal bodies. After samples have been analyzed, the data help build a picture about the spread of pollution and the subpopulations most likely to retain high levels of chemicals.

or chemistry—had 165 chemicals in his body out of the 320 chemicals tested. Duncan wrote, "I set out to learn, as best I could, where those toxic traces came from. Some of them date back to my time in the womb, when my mother downloaded part of her own chemical burden through the placenta and the umbilical cord. More came after I was born, in her breast milk. Once weaned, I began collecting my own chemicals as I grew up. . . ." Others in the public may possibly have a chemical profile similar to the writer's and summarized in the following table. Most of these chemicals appear in people in a parts per billion (ppb) concentration. To learn about possible chemicals in the body, anyone can access the CDC's annual *National Report on Human Exposure to Environmental Chemicals* that contains biomonitoring results from U.S. residents.

In the United States, two government agencies conduct biomonitoring: the EPA and the CDC. The EPA manages the National Human Monitoring Program and the Voluntary Children's Chemical Evaluation Program

Example of Chemicals Found in an Adult Body (Blood and Urine Only)

Chemical	Number Tested	Number Found
polychlorinated biphenyls (PCBs)	209	97
polybrominated diphenyl ethers (PBDEs)	40	25
pesticides	28	16
dioxins	17	10
phthalates	7	7
perfluoroalkoxy fluorocarbon (PFAs)	13	7
metals	4	3
bisphenols	2	0

(VCCEP). The CDC runs the National Health and Nutrition Examination Survey (NHANES) and the National Human Exposure Assessment Survey (NHEXAS). These federal agencies have also joined with other industry biomonitoring programs to measure chemicals in farmworkers and workers in pesticide manufacture. Internationally, three organizations conduct large biomonitoring projects: the European Union's Science, Children, Awareness, Legislation and Evaluation (SCALE) initiative; Canada's Domestic Substances List; and the Organization for Economic Cooperation and Development.

Chemicals released into the environment by accident pose a health threat because they come in sudden large doses, perhaps at levels higher than the human body can withstand. The following sidebar, "The Bhopal Accident," examines one such chemical accident.

THE BHOPAL ACCIDENT

On December 2, 1984, an environmental disaster unfolded in the city of Bhopal in central India. An underground and corroded chemical storage tank had exploded at Union Carbide's pesticide manufacturing plant. Highly toxic methyl isocyanate gas drifted into the city, killing at least 3,800 people—unofficial estimates reached as high as 10,000—and injuring thousands more. The toxic cloud quickly enveloped a 23-square-mile (60 km^2) area and instantly killed more than 2,000 people, then drifted, causing additional fatalities. Many survivors experienced violent coughing, swelling of the lungs, neurological damage, bleeding, and impaired vision and blindness. By the time the cloud had extended to 30 square miles (78 km^2), 600,000 people had been exposed to it. In the year 2001, the International Campaign for Justice in Bhopal estimated the death toll due to the accident had reached 20,000 and chronic illnesses had afflicted at least 150,000. Children of parents who were exposed to the chemical, and subsequent generations, have suffered birth defects. In 2008 the United Kingdom's *Guardian* reported, "Medical experts who studied the effects of the gas on children born in communities affected by the gas cloud said there was now 'no doubt of increased chance of the negative effects in children'." Indian doctors will likely continue to see the effects of the accident for years.

The Bhopal incident involved five errors that set the disaster in motion. First, a water leak had caused the storage tank to corrode and no one had been monitoring the site for this or any other type of damage. Second, Bhopal had no emergency evacuation plan for its residents. Third, the city lacked a warning system and the plant's early warning siren had been shut off to save money.

CLASSIFYING TOXICITY

LD_{50} values allow scientists to compare the toxicity of all animals on the same scale, and then rank each chemical according to its toxicity level. No chemical achieves a rating of "nontoxic" because every substance can be lethal, even though the lethal dose may be very large. Examples of high-toxicity to low-toxicity substances are shown in the table on page 62.

The toxicity rating of every industrial chemical would require an enormous amount of testing, so environmental chemicals can choose to view all chemicals of unknown toxicity in one of two ways. In one approach, chemicals are assumed to be safe until science amasses evidence to show that the chemical is dangerous at low doses. Conversely, scientists might

Fourth, the plant's safety equipment did not go into automatic shutdown mode to stop gas leaks. In fact, the manufacturer had taken shortcuts in its safety and maintenance programs to save money. Fifth, a tent city of homeless lived within the shadow of the facility and likely had no idea of the dangers that were close to them.

In Bhopal's aftermath, various nations set up programs to make safety and emergency procedures part of their chemical industries. In the United States in 1986, Congress passed the Emergency Planning and Community Right-to-Know Act for the purpose of making public all information on chemical hazards in a community. India's Central Bureau of Investigation meanwhile traced the events that led to the accident in order to construct a better safety system.

Despite these actions, no one has been held accountable for the accident. In 1992 a Bhopal court found Union Carbide Chief Executive Officer Warren Anderson responsible for the event. Over a decade later, however, the environmental group Greenpeace noted, "Anderson has been hiding in the United States since an explosion at his company's plant in Bhopal caused the immediate deaths of thousands of people and led to life-long suffering for almost 120,000 survivors. He is wanted in India to face charges of culpable homicide over the deaths of 20,000 people since the disaster." Little progress has been made in bringing anyone to justice for the disaster.

The abandoned site in Bhopal still contains about 8,900 tons (8,000 metric tons) of pesticides that continue to leak from the site with each rain. A speedy cleanup for Bhopal is no longer possible, but the lessons from this disaster should remain forever.

Toxicity Rating for Humans (155 pound [70 kg] body weight)				
Category	Rating	Dose, LD_{50}	Example	Label Warning
6	super toxic	less than 5 mg/kg	nerve gas	DANGER POISON
5	extremely toxic	5–50 mg/kg	potassium cyanide	DANGER POISON
4	very toxic	50–500 mg/kg	mercury salts	WARNING
3	moderately toxic	0.5–5 g/kg	phenobarbital	CAUTION
2	slightly toxic	5–15 g/kg	ethyl alcohol	not required
1	almost nontoxic	more than 15 g/kg	glycerin	not required

choose to consider all unranked chemicals to be toxic because there no evidence exists to prove otherwise.

Chemicals of known toxicity can have two effects on the body: *direct effects* or *indirect effects.* Direct effects consist of symptoms that occur as a result of ingesting, inhaling, or absorbing a toxin. Changes in the body's physiology or in reproduction, or coma and death represent direct effects. Indirect effects consist of damages to health even if a person has not come in contact with a toxin. For example, a birth defect in a newborn due to the mother's exposure to mercury is an indirect effect.

The EPA's and CDC's information on chemicals help the public understand the potential dangers of chemicals. The sidebar "The Material Safety Data Sheet" explains an important resource for people who work with chemicals.

POINT AND NONPOINT POLLUTION SOURCES

An important part of an epidemiology comprises finding a toxin's source. Environmental toxins come either from *point sources,* which emit pollution from a single location, or *nonpoint sources,* which emit pollution from many dispersed locations. Examples of point sources are pipes, wells, storage canals, waste lagoons, dump sites, landfills, ditches, sewers, faulty wastewater treatment plants, and smokestacks. Examples of nonpoint sources may be runoff, eroded soils and sediments, agricultural wastes, pollution carried downstream, pesticide spraying, lawns, vehicle emissions, and natural events such as floods, storms, or volcanoes. Point and nonpoint sources can also be stationary things, such as smokestacks, or moving objects, such as trucks.

Point sources often lead to a single exposure to a toxin, or repeated single exposures. Nonpoint sources, by contrast, have the potential of causing chronic exposure because they can be difficult to identify and control. Meanwhile, the nonpoint sources continue to pollute and expose people to toxins. Nonpoint sources are also more expensive to control than point sources because they may involve more than one industry and more than one community. Exposure to either type of source can be hazardous; chemical toxicities correlate with the type of chemical to which a person is exposed, not to the chemical's source.

The Material Safety Data Sheet (MSDS)

An MSDS is a document published by a chemical manufacturer to state the dangers of a chemical and provide information on its safe handling. The Occupational Safety and Health Administration (OSHA) requires that the following groups have an MSDS available to them for each chemical at their job site: (1) employees who handle chemicals; (2) employers who need information on the proper safe storage of a chemical; and (3) emergency responders who enter a site that contains chemicals. Emergency responders who consult an MSDS if they expect to come in contact with a dangerous chemical are firefighters, paramedics, hazardous materials crews, emergency room staff, and poison control centers. The following table describes the information on an MSDS.

Elements of a Material Safety Data Sheet

Section	Required Information	Support Information
heading	substance's name as it appears on the label	other nomenclature used for the chemical
I	manufacturer's name and contact information	name, address, telephone, emergency telephone, date of the MSDS
II	hazardous ingredients/ identity information	list of all hazardous components that make up the substance, including exposure limits

HOW TOXINS ENTER THE BODY

People take in chemical or biological toxins through one of three entry mechanisms: (1) oral uptake by ingesting food or water; (2) respiratory uptake by inhaling air; or (3) dermal or absorption uptake whereby the skin (or eyes) comes in contact with a toxin. Toxins ingested with food or water first contact the mouth and then the esophagus of the upper gastrointestinal tract. The stomach absorbs only a small amount of nutrients and so probably absorbs little toxin. The small intestines and the large intestine, or colon, possess a strong capacity to absorb com-

Section	Required Information	Support Information
III	physical and chemical characteristics	boiling point, vapor pressure, vapor density, specific gravity, melting point, evaporation rate, solubility in water, appearance, and odor
IV	fire and explosion hazards	flashpoint, flammability limits, extinguishing media, special firefighting procedures, and unusual fire and explosion hazards
V	reactivity	stability, conditions to avoid, incompatible materials, decomposition by-products, and conditions for hazardous polymerization
VI	health hazards	routes of entry into the body, health hazards, carcinogenicity, signs and symptoms of exposure, medical conditions aggravated by exposure, and emergency and first aid procedures
VII	precautions for safe handling and use	procedures for spills, disposal, handling, and storage
VIII	control measures	respiratory protection, gloves, eye protection, and other protective clothing or equipment

pounds, so most ingested toxins likely enter the bloodstream through these organs.

Toxins that have been inhaled enter the lungs after passing through the nasal canals, pharynx, larynx, and the trachea. Bronchial tubes branch off from the trachea and carry air to the alveoli. Alveoli are the final branches, which provide a transfer point at which oxygen in the inhaled air moves across a barrier to the blood. Very small vessels called pulmonary capillaries accept the oxygen at the vessels' gas-blood barrier, and then the capillaries carry the oxygen to larger vessels, which supply it to tissues. Inhaled toxins may take one of two routes in the

Environmental scientists and government agencies can control point sources of pollution much more easily than they can control most nonpoint sources. For example, officials can monitor a point source such as the Gösgen-Däniken nuclear power plant in Switzerland for radioactive emissions, electromagnetic emissions, and hot water released into the Aare River, shown in the foreground. *(NAGRA)*

scheme: (1) by entering the bloodstream at the gas-blood barrier or (2) by lodging in lung tissue.

Dermal absorption depends on the type of chemical that contacts the skin. For instance, the skin absorbs alcohols and organic solvents differently than it absorbs chemicals dissolved in water. Substances move through the skin layers by diffusion, a transport process that uses no energy. Once substances cross the skin layer called the dermis, capillaries or lymph vessels take the substances and carry them throughout the body.

The bloodstream carries many toxins directly to the liver where they become detoxified, or made less hazardous. The liver's enzyme systems detoxify virtually every substance that is harmful to the body: metabolic wastes (such as ammonia), alcohol, drugs, and environmental toxins. These enzymes are collectively called *mixed function oxidases* because they use oxygen as part of the reaction that converts two different compounds (mixed function) to a different pair of compounds, shown in the following equation:

$$AH + BH_2 + O_2 \rightarrow AOH + B + H_2O$$

The liver protects the body using mixed function reactions by converting toxic substances to less toxic forms that the body then eliminates. In a few instances, however, large synthetic compounds may be converted to a version more toxic than the original structure. A second hazard arises from a process called *synergism,* in which a natural compound in the body blocks oxidase reactions and so increases a chemical's toxicity. Some pesticides, for instance, cause synergistic reactions in the liver.

Detoxified wastes leave the body through the kidneys, colon, lungs, or skin. Lactating females also eliminate some detoxified compounds in breast milk, which can be transferred to an infant. The kidneys excrete chemicals from the bloodstream by filtering the blood and eliminating wastes in a process called *selective resorption.* Selective resorption in the kidneys serves an important role in the body by recovering glucose, amino acids, vitamins, calcium, and potassium, but excreting toxic wastes.

The body rids itself of many water-soluble toxins fairly easily but it has a harder time eliminating persistent toxins that accumulate in fatty tissues. Of the major environmental toxins, pesticides, organic compounds, and heavy metals receive the most attention because of their effects on blood, tissues, or organs.

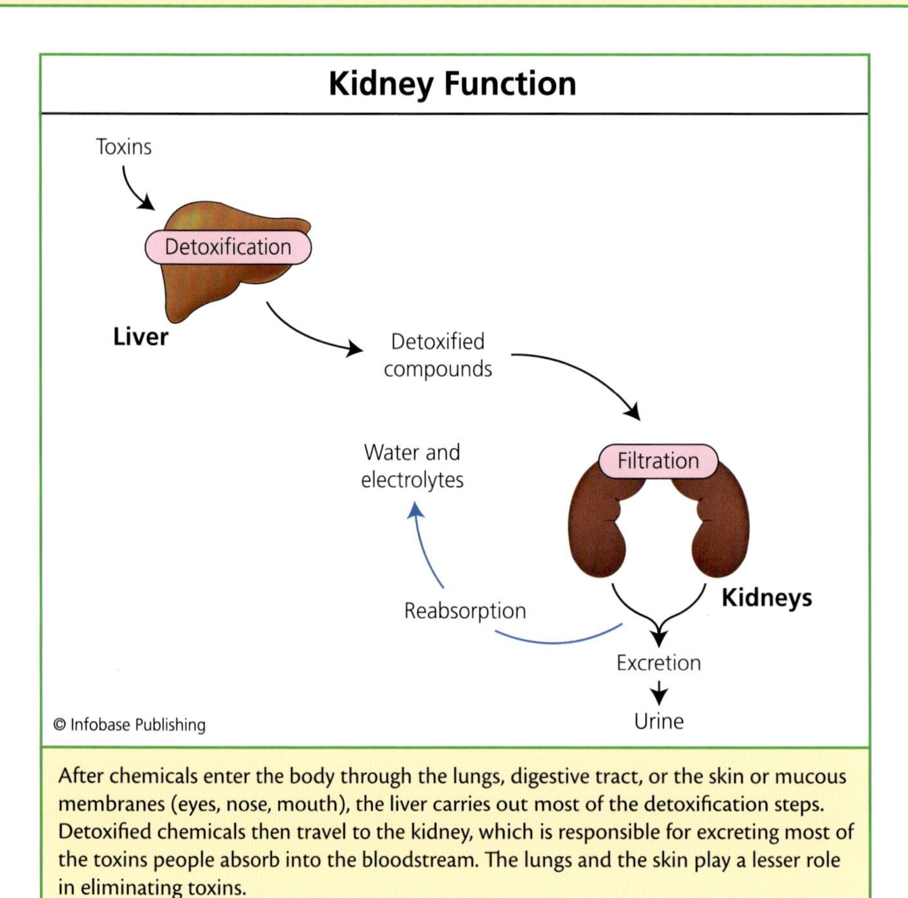

After chemicals enter the body through the lungs, digestive tract, or the skin or mucous membranes (eyes, nose, mouth), the liver carries out most of the detoxification steps. Detoxified chemicals then travel to the kidney, which is responsible for excreting most of the toxins people absorb into the bloodstream. The lungs and the skin play a lesser role in eliminating toxins.

PESTICIDES

Pesticides are compounds designed to control animal and plant life that endanger agricultural crops or gardens. The four main pesticide categories are the following: (1) insecticides, which control insects; (2) herbicides, which control weed growth; (3) fungicides, which control fungi; and (4) rodenticides, which kill rodents. The EPA regulates the manufacture and use of pesticides under a law called the Federal Insecticide, Fungicide and Rodenticide Act (FIFRA). FIFRA became law in 1947 but the EPA acquired responsibility for enforcing it in 1970 for the purpose of ensuring the safety of all pesticides in people, animals, and aquatic life.

Pesticides enter the environment by the following routes: spray pesticides move in the air and eventually end up in water or soils, and pesticides applied directly to soils filter into groundwaters or wash off into surface waters. Paint preservatives and boat protectants also add to the total amount of pesticides in the environment, and all of these chemicals move remarkable distances around the globe.

In 1967 British chemists J. O'G. Tatton and J. H. A. Ruzicka reported in *Nature* magazine the discovery of pesticides in Antarctica's wildlife. At first, they assumed the chemicals came from an Antarctic research station, but further studies implied the pesticides had come from much farther away. Since Tatton and Ruzicka's studies, scientists have found an alarming array of pesticides, such as chlordane, in penguin colonies. Martin Riddle of the Australian Antarctic Division told Australia's ABC News in 2008, "Some of these chemicals were obtained to 120 times the levels that were in other parts of the Antarctic." Though dichlorodiphenyltrichloroethane (DDT) levels are slowly dropping in wildlife across the world, the Adélie penguins seem to be holding onto DDT in their fat. In 2008 the *New York Times* reported, "DDT levels in these birds have remained about the same in the past thirty years, and the researchers say it's likely that Antarctic glaciers, which would have accumulated DDT through [the atmosphere], are delivering the pesticide into the food chain through meltwater." Riddle explained, "What this demonstrated is that penguins were actually biomagnifying . . . and they are creating small hot spots around their colonies." Pesticides may well have entered populations of other animals, including endangered species.

Some places in the United States continue to hold amounts of DDT almost 40 years after the U.S. ban on the use of this pesticide. Offshore from Los Angeles the Pacific Ocean contains a 100-ton (91–metric ton) deposit of DDT spread over several square miles. "Things have not changed a whole lot in the last decade or so," said fish biologist David Witting of the National Atmospheric and Oceanic Administration to the *Los Angeles Times* in 2007. This problem occurs for two reasons. First, DDT persists in the environment for a long time, and second, other countries continue to use this pesticide.

Pesticides cause a variety of ills in humans, fish, and wildlife, mainly reproductive disorders, endocrine disruption (see the "Endocrine Disrupters" sidebar on page 72), immune system problems, nervous system impairment, and cancer. DDT and its breakdown product dichlorodiphenylethylene (DDE) both harm bodily systems. In the 1960s DDE was identified as the cause of weakened eggshells in eagles, falcons, and other birds. Only the DDT ban coupled with protections under the Endangered

Pesticide and herbicide applications have been a major source of environmental chemicals for decades in the United States and other countries. Pesticides kill animal life such as insects; herbicides kill weeds. Global pesticide and herbicide use exceeds 5 billion pounds (2.3 billion kg) a year. The United States uses about 25 percent of that total. *(Californians for Pesticide Reform)*

Species Act of 1972 saved these species. Pollution expert Mariann Lloyd-Smith told ABC News in 2008, "The problem is once you have a contaminant like that in the environment and then it makes its way into the food chain, ultimately it makes its way into the human food chain." Individuals have little power against this type of pollution. Only national governments and international organizations have the capacity to regulate the chemicals that enter the environment. Environmentalists encourage the enforcement of more bans against hazardous chemicals, but these changes take time to make their way through government and industry.

ORGANIC COMPOUNDS AND SOLVENTS

Most pesticides are also organic compounds, which are a large group of environmental contaminants also found in plastics, paints, finishes, industrial solvents, dyes and inks, flame retardants, coatings, and textiles. It would be almost impossible to walk through a typical home in the industrialized world and not touch an item containing organic compounds, some of which are quite toxic.

A 2005 CDC report titled *Third National Report on Human Exposure to Environmental Chemicals* lists a bewildering array of organic compounds that have been detected in people's bodies. *National Geographic* stated in 2006, "This is the dark side of industrial chemistry, which gives us convenient products and abundant food but exacts a human cost." Appendix D provides a list of non-pesticide organic compounds and organic solvents that today cause concern in environmental medicine. Some of these are highlighted in the Sidebar "Case Study: San Francisco's Ban on Toxic Toys" on page 78.

Industries use or produce organic chemicals and solvents in the synthesis of other chemicals, as by-products of chemical reactions, or in the cleaning of equipment. Organic compounds also occur in products used by people every day. For instance, the Organic Consumers Association announced in 2008 that it had detected the petroleum-based solvent 1,4-dioxane in 46 of 100 personal care products it tested, but the U.S. Food and Drug Administration (FDA) stated that dioxane-containing products

Pelona Schist is a type of greenish rock that forms outcroppings in California's San Gabriel Mountains. This rock is a good source of iron and magnesium, but it also contains chromium, which the U.S. Geological Survey has determined contaminates the groundwater in parts of southern California. The schist gives an example of a natural source of toxic contamination of the environment. *(USGS)*

ENDOCRINE DISRUPTERS

Endocrine disrupters are also known as hormone disrupters or hormonally active agents (HAAs). When these organic compounds enter the body by ingestion or breathing polluted air, they interfere with the body's system for turning on or off hormone production. Endocrine disrupters are thought to do this by mimicking the activity of natural hormones. As a consequence, endocrine disrupters upset sexual reproduction, growth, body development, learning capacity, and behavior. Researchers have conducted a small number of studies on these compounds and the results suggest endocrine disrupters may cause improper development of sexual organs in baby boys, reproductive-tract cancers, lowered sperm production, and factors leading to diabetes.

The following compounds have been suspected of acting as endocrine disrupters: some pharmaceuticals, dioxins, PCBs, hexachlorobenzene, some flame retardants, phthalates, DDT and other pesticides, bisphenol A, ingredients in detergents and cosmetics, and materials in children's toys. Wastewater treatment may not remove all of the endocrine disrupters in wastewater, so these compounds could return to the environment in the treatment plant's outflow.

The first alarm regarding endocrine disrupters came from discoveries of unusual aquatic life that had been exposed to contaminated runoff. In 2007 the Associated Press reported one such

"do not present a hazard to consumers." Other organic compounds can be found in common products, but their concentrations are very low and they have not been associated with health problems in people. California has nonetheless taken the lead in controlling the amounts of chemicals such as dioxane in products; the state requires a warning label on any product containing more than 30 parts per million (ppm) dioxane.

HEAVY METALS

Metals make up the elements in the periodic table of chemicals that have a positive charge, usually conduct electrical current, and possess a shiny luster. Heavy metals make up a loosely defined subgroup of dense metals that are toxic to living tissues at low concentrations. The following metals are heavy metals: cadmium, chromium, copper, lead, mercury, thallium, and zinc. Some texts also include iron, manganese, molybdenum, and cobalt as heavy metals, summarized in the table on page 74. Of these metals, zinc and cobalt are essential nutrients in

example in the Potomac River near Washington, D.C. Since 2003, biologists have found in the Potomac intersex fish, male fish that produce eggs in their testes. "Endocrine disrupters . . . may contribute to the high percentage of male smallmouth bass in the Potomac that exhibit female characteristics," said a spokesperson from the U.S. Geological Survey. "We analyzed samples of 30 smallmouth bass from six sites, including male and female fish without intersex and male fish with intersex. All samples contained detectable levels of at least one known endocrine-disrupting compound, including samples from fish without intersex."

Not surprisingly, endocrine disrupters have generated their share of controversy between scientists and chemical companies. In 2007 *USA Today* quoted a representative from the American Chemistry Council as saying, "There is no reliable evidence that any phthalate has ever caused a health problem for a human from its intended use. Some organizations have 'cherry picked' the results showing impacts on test animals to create unwarranted concern." Like hundreds of other chemicals that enter the environment, much more testing should be done on endocrine disrupters for the purpose of developing a clear picture of their effects on human health. The disquieting effects that these compounds have on animals certainly make endocrine disrupters a priority in environmental medicine.

humans, cobalt as a component of vitamin B_{12}. The recommended dose of zinc for males 14 and older is 11 mg daily (9 mg for women 14–18, 8 mg for 19+); Cobalt needs are met with a daily intake of 2.4 mg of vitamin B_{12}.

Heavy metals bioaccumulate in food chains because neither plants nor animals degrade these elements or convert them into safer forms. Humans at the top of food chains receive the highest doses of several heavy metals, most notably mercury. Heavy metals may be ingested with food or water, inhaled, or absorbed through the skin.

A devastating mercury poisoning event from the 1950s through the 1970s highlighted the dangers of heavy metals. In the 1950s a plastics factory in Minamata on the southern tip of Japan discharged large amounts of mercury into Minamata Bay. Bacteria converted the metal to methyl mercury, a form that concentrates in living tissue. Methyl mercury accumulated in Minamata Bay's food chains and many hundreds of people, wildlife, and pets eventually ingested large amounts. At least 50 people died; more than 100 people suffered permanent disabilities. Most

HEAVY METALS		
METAL	**SOURCES**	**MAIN TOXICITY OUTCOMES**
arsenic	semiconductors, glass, wood preservatives	gastrointestinal and lung irritations, decreased red and white blood cells, cancer, death
cadmium	electroplating, batteries, nuclear reactor control rods, stabilizers for plastics, tobacco smoke	lung damage, bone fracture, kidney failure, infertility, damage to nervous and immune systems, death
chromium	coating on other metals, stainless steel, pigments in cements and plaster, leather tanning, gems, recording tapes	allergy, rash, nose irritation, ulcers, respiratory problems, weakened immunity, kidney and liver damage, lung cancer, death
cobalt	alloys, magnets, paints, inks, pottery, stained glass, tile, jewelry	respiratory, heart, thyroid, and vision damage
copper	electrical equipment, plumbing, heat exchangers, alloys	neurological effects, anemia (liver cirrhosis, brain damage, kidney failure, and copper deposition in cornea)
iron	mining, blast furnaces in steel manufacture, cookware vehicles, appliances	respiratory siderosis (a lung disease caused by inhaling iron compounds), possible eye damage, possible cancer
lead	batteries, television and computer screens, pipes, alloys, paint, gasoline, crystals, nuclear reaction shielding, tobacco smoke	increased blood pressure, brain and nerve damage, miscarriage, decreased fertility, behavioral disorders, learning disabilities in children
manganese	iron and steel production	brain disorders, nerve damage

Metal	Sources	Main Toxicity Outcomes
mercury	burning fuels, incinerators, industrial emissions, mining, paper or steel manufacture, electronics	weakness, loss of peripheral vision, tremors, behavioral disorders, coma, death
molybdenum	alloys, electrodes, catalysts	liver failure
thallium	lenses, photocells, thermometers	irreversible nerve damage
zinc	galvanized iron and steel, alloys, pigments	metal hypersensitivity, pancreas damage

disturbing, a generation of children suffered from serious birth defects and deformities. *Minamata disease* has since been found to occur elsewhere and it likely kills a large number of wildlife and birds.

Heavy metals cause the most serious metal-related illnesses, but other metals in the environment also harm people and wildlife. The table on page 76 lists metals that cause additional health concerns.

Many heavy metals have been confirmed or suspected of causing cancer when people have been exposed to low chronic but levels. Acute toxicity, by contrast, arises from a single high dose of toxic metal, most often affects the nervous system, and may be a special concern to people who work in metal industries—businesses that make metal raw materials (sheet metals, rods, piping, etc.) or businesses that make metal products.

TISSUE AND CELL FUNCTION

The damage that most toxic chemicals do to tissue can be reversed if a person avoids further exposure and the body has time to eliminate the chemical and repair the damage. Heavy metals are an exception: They cause permanent damage in living tissue because the body has a difficult time excreting even trace amounts of metal. People therefore can protect their

Metal Toxicities		
Metal	**Sources**	**Main Toxicity Outcomes**
aluminum	household utensils, containers, alloys, soil	possible nerve damage
beryllium	alloys, semiconductors, microwave devices, satellite optical systems, nuclear reactors	lung diseases, anorexia and weight loss, rapid heart rate, death
gold	mining, coatings, alloys, electrical switches and connectors, jewelry	skin, eye, and respiratory irritations
nickel	magnets, alloys, platings, glass tints	possible lung cancer or nasal tumors, skin irritation, allergy
platinum	aircraft electrical parts, catalytic converters, glass, jewelry, liquid crystal displays	DNA damage, allergy, kidney and bone damage, hearing impairment, cancer
tungsten	incandescent lightbulbs, steel alloys	skin, eye, and lung irritations
yttrium	televisions, fluorescent lamps, lasers	lung embolisms, liver damage, cancer

health by immediately stopping any known exposure to toxic chemicals and by avoiding any exposure to heavy metals such as mercury, lead, and cadmium. Workers who must be exposed to these metals in their jobs should use the proper safety equipment, such as protective clothing and breathing devices.

Environmental toxins in the body tend to change the speed of the body's functions—increased heart rate, increased sweating, hyperactivity, and altered breathing rate. These changes come about because of two main factors in toxicity: the chemical's LD_{50} and the exposure route, meaning the route that the toxin takes to get to a certain organ. Therefore, toxicity can be expressed either as acute versus chronic, as already described, or as an organ toxicity. For example, renal toxicity and liver toxicity mean that most of the damage takes place in the kidneys and the liver, respectively.

Environmental medicine uses a method called *bioassay* to determine the effect of a toxin on living organisms, tissues, or cells. Bioassays include use at least two groups of subjects: one group exposed to the potential toxic chemical and another group (called a control group) that is exposed only to water. Bioassays on tissue or cells, rather than organisms, use a

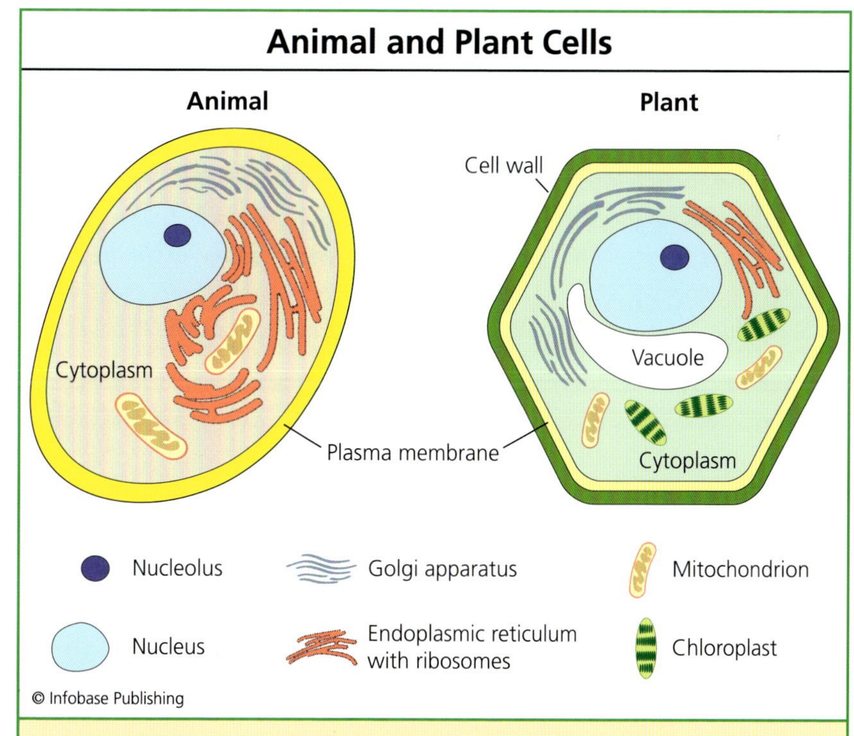

The study of toxins depends on knowledge of cell structure, including the methods of cell reproduction and nutrient (or toxic chemical) uptake. Animal cells do not possess a strong protective cell wall, so they may be more vulnerable to the entry of toxins than plant cells. Once inside an animal cell, the toxin disrupts many of the mechanisms that keep the cell functioning. These mechanisms include reproduction starting in the nucleus, protein production at the endoplasmic reticulum, and energy generation in the mitochondria.

CASE STUDY: SAN FRANCISCO'S BAN ON TOXIC TOYS

Even if all environmental toxins were to be banned this minute, hundreds of them would persist in the environment and in many consumer products. San Francisco, California, has tried to stem the exposure of people to dangerous substances by banning certain chemicals from products. On December 1, 2006, San Francisco banned the sale, distribution, and manufacture of children's (younger than three years) products containing bisphenol A and high levels of phthalates. But six months later the city repealed the ban because of a lack of evidence on the dangers of these compounds. The ban and its repeal highlight the confusion that arises over discussion of chemicals and their safety when they occur at very low levels.

The plastics industry uses bisphenol A as a component in hard, clear materials, but it has also been found in adults and fetuses. Researchers have gathered evidence that low levels of bisphenol A affect the thyroid gland, brain, pancreas, and prostate gland functioning. These latter studies have been done in laboratory experiments, however, and not on humans, so the debate over bisphenol A's safety continues.

Scientists have taken opposing sides in the argument over chemical bans, partly explaining why San Francisco instituted its ban and then reversed itself. At the time of the ban, chemical industry spokesperson Steve Hentges assured, "In every case, the weight of evidence supports the conclusion that bisphenol A is not a risk to human health at extremely low levels to which people might be exposed." Parents and other consumers were left to make their own decisions about the products they bought. San Francisco Bay Area resident and mother Mary Brune told the *San Francisco Chronicle* in 2006: "It's impossible to keep plastic toys out of children's mouth[s]. They

technique called *tissue culture* in which scientists grow cells grow in a nutrient solution inside an incubator.

Toxins work at three different levels in the body: extracellular, intracellular, and nuclear. Extracellular activity occurs when a toxin affects the function of an entire organ. For example, endocrine disrupters act on endocrine glands as if the disrupters themselves are hormones, while beryllium affects the heart by increasing heart rate. Intracellular toxins affect the cell's enzymes, membranes, or cell components other than the nucleus. Nuclear toxins target things inside the nucleus: genes, DNA activities, RNA, or meiosis, the process in which cells divide and split their genetic material (the chromosome) in half. Because intracellular and nuclear toxins target cells, this type of toxicity is called *cytotoxicity*.

chew on things. So we as parents rely on the manufacturers of products to ensure their safety. If consumers demand safer products and businesses demand safer products from their suppliers, we'll be able to get these toxic products off our shelves." Brune's straightforward idea sums up the best approach to controlling harmful chemicals.

Demanding safer products as Brune suggests is a good tactic but a difficult one. Environmental health professor Joel Tickner explained for the *Chronicle* why he holds little faith in current safety testing methods: "Protections for children from chemicals in toys are weak at best and dysfunctional at worst. Consumers would be astonished if they knew that federal laws regulating chemicals in children's toys all require balancing the benefits of protecting children with the costs to industry of [offering] safer alternatives." If what Tickner says is correct, government control of chemicals must be improved in order to protect people's health.

By 2008 new studies conducted at the National Institutes of Health in Washington, D.C., prompted scientists to express "some concern" that bisphenol A was a health hazard to humans. Hentges continued to assure the public that "those low levels [in plastics] are not a health concern" and he may, in fact, have a point: The dangers from a chemical relate to the chemical's dose.

San Francisco and the rest of the United States continue to monitor worldwide opinion on chemicals that leach out of plastics. The European Union allows a daily human exposure to bisphenol A of up to 9/100,000 of an ounce per pound of body weight (5 mg/kg). Meanwhile, armed with what it believes is new and compelling evidence against plastics, the California Senate introduced in 2008 a new ban against *plasticizers* at any level detected in baby products.

Extracellular toxins may interfere with the way the body's tissues and cells communicate with each other by hormones or chemical signals. In 2006 University of Cincinnati researcher Michael Borchers explained how he believes toxins affect lung tissue: "When tissue is exposed to a pathogen (disease-causing agent), the immune system immediately wants to destroy the damaged cells so healthy tissue can take over. But when the lungs experience chronic, low-level damage, we believe at some point that damage exceeds the body's natural ability to repair tissue, and through the destruction of lung tissue, it may actually start contributing to chronic lung disease instead of protecting against it." If Borchers is correct, for each toxin that enters the human body, a similar battle ensues between the effects of the toxin and the defenses of organs and cells.

Inside cells, some toxins simply accumulate to higher and higher levels until a cell cannot carry out its normal jobs of absorbing nutrients from the blood and reproducing. As a result, a person may feel fatigue or weakness. Other toxins act in a more specific manner by disrupting membrane components such as enzyme systems, nutrient uptake, or energy reactions. One type of cellular reaction called *oxidation* causes the formation of *free radicals,* which are highly reactive chemicals that cause havoc inside cells. Biochemistry researcher Jeffery Blumberg of Tufts University explained to the *Egyptian Doctor's Guide* in 2006, "If free radicals simply killed a cell, it wouldn't be so bad . . . the body could just regenerate another one. The problem is, free radicals often injure the cell, damaging the DNA, which creates the seed for disease."

The liver protects itself and the rest of the body from toxic damage by acting as the body's main detoxification center, converting toxic compounds to less toxic forms that the body can excrete. The liver repairs its own cells injured by toxins or it grows new cells, and by these activities healthy people receive protection from many hazardous substances. During chronic exposure to hazardous substances, the liver undergoes constant repair, which causes scar tissue to build up in the liver. In the worst cases, the liver becomes inflamed and cannot detoxify any more toxins. Chronic exposure therefore reduces the liver's ability to protect the body from poisons.

Neurotoxins attack nerve cells and poison the liver, and they threaten the body with permanent or very long-term nerve damage. Neurotoxins belong to three general groups, as follows:

- Type 1 neurotoxins produce mild disorders such as irritability, fatigue, memory problems, and changes in mood.
- Type 2 neurotoxins produce two levels of chronic disorders: Type 2A neurotoxins cause long-term personality changes and type 2B neurotoxins cause problems with intellectual functions (decreased concentration, memory loss, and trouble learning).
- Type 3 neurotoxins produce irreversible dementia and deterioration of brain functions.

Many neurotoxins interfere with two nerve cell components: membrane proteins and the ion channels that transport molecules into or out of the cell. Ion channels are porelike protein structures that move charged molecules through the cell membrane. Toxins that interfere with these channels disrupt the normal communication between nerve cells and that of nerve cells with muscle. Organic solvents work differently by attacking the fatty components inside cell membranes. By attacking nerve-cell membranes in this way, organic solvents impair the central nervous system (brain and spinal column) as well as the peripheral nervous system (nerves of the trunk and extremities).

Radioactive substances cause wide-ranging health effects throughout the body, summarized in the following table.

GENE FUNCTION

In 2005 medical researchers found a link between pesticides and the occurrence of Parkinson's disease. Since then Duke University geneticist Randy Jirtle has lead research on a new area of gene control called *epigenetics*. An *epigenome* may be thought of as an instruction set for all of the genes in the body; it tells which genes should turn on and which

Radioactivity Effects on the Body

Organ or Function	Health Effects
thyroid gland	absorbed iodine 131 damages tissue with *gamma radiation*
sight	radiation causes the lens to opacify (become cloudy)
cardiovascular disease	high radiation doses injure muscular function, heart valves, and electric conduction; leads to myocardial infarction, pericarditis, atherosclerosis
immune system	damage to T-lymphocytes, interleukin; abnormal levels of immunoglobulins
reproduction	decreased or zero sperm counts, temporary ovary damage, damage to fetus development
nervous system	damage to brain cells at high radiation doses

ones should turn off, and when. The epigenome helps bodies adapt to immediate changes in the environment without using mutations for adaptation.

A writer for the university's science newsletter, *The Green Grok,* explained Jirtle's work in 2008: "For quite some time scientists have been trying to determine how exposure to environmental toxins can result in serious disease years or even decades later. Epigenetics may provide the mechanism. An exposure to an environmental toxin at one point in a person's life (and most critically during gestation) can trigger the epigenome to turn on or turn off a key gene. Years later, because of that epigenetic change, a disease may appear."

Environmental science does not have all the answers yet as to how environmental toxins interfere with *gene expression.* Michael Skinner, director of the Center of Reproductive Biology at Washington State University, suspected a relationship between environmental toxins and genetic disease in a September 25, 2006, issue of *Medical News Today.* "Only the original generation mother was exposed to an environmental toxicant," Skinner explained, describing an animal study. "A human analogy would be if your grandmother was exposed to an environmental toxicant during mid-gestation, you may develop a disease state even though you never had direct exposure, and you may pass it on to your great-grandchildren." If this theory proves true, it will highlight the ability of environmental toxins to enter populations and affect health for generations.

The association between environmental toxins and birth defects still holds many questions, but this association seems to follow these generalizations:

- Pesticides cause musculoskeletal system defects.
- Metals and solvents cause nervous system defects.
- Plastics cause chromosomal defects.

NATURAL DETOXIFICATION

The liver is the body's largest organ and accepts about a gallon (3.8 l) of blood every 2.6 minutes, which it filters through a complex network of arteries, capillaries, and veins. Enzymes inside liver cells, or hepatocytes, rid the body of toxins by a variety of processes. Many environmental toxins dissolve in fatty materials and the liver's enzymes convert these fat-soluble compounds into water-soluble compounds that the blood carries away to the colon or kidneys for excretion. Some liver enzymes break down the toxin into smaller nontoxic molecules, ready for excretion.

The liver converts the body's wastes and toxic compounds into less harmful compounds in an enzymatic process called *biotransformation.* The main group of liver enzymes that biotransform toxins belong to a family of enzymes known as cytochrome P_{450} enzymes, so called because they all depend on the protein cytochrome P_{450} to help them carry out their reactions. Cytochrome P_{450} enzymes detoxify mainly dioxins, phosphorus-containing pesticides, and air pollutants.

Vitamins B, C, and E and the element selenium aid liver detoxification enzymes, and because of this some people think that diet alone may detoxify poisons. Any foods that supply these nutrients certainly help the liver work at its best, but Western medicine has not emphasized detoxification through diet. Eastern medicines and holistic healing (medical care by treating organ systems and the mind) have put greater emphasis on detoxification by dietary means. At this time, doctors do not agree on the extent to which diet alone can combat the effects of toxic exposure.

TOXIN POISONING THERAPY

Different toxin poisonings require various treatments. The treatment that doctors choose for a person who suffers from a toxicity takes into account the dose, the age of the person, and any existing health conditions. Organic solvents offer the best chances for a complete recovery because a person can excrete them with normal exhaling. If exposure has ended, organic solvents can leave the body within days or weeks. People who suffer chronic exposure to solvents or those in high-risk health conditions might require the use of a respiratory support machine or physical therapy to increase heart rate and blood flow. Doctors may also choose to prescribe antidepressant drugs for high-level exposures because organic solvents can lead to depression and other behavioral changes. Overall, however, organic solvent poisoning presents an easier treatment than pesticide or heavy metal toxicity.

Pesticide therapy usually begins with decontaminating any exposed skin by repeated showering with mild soap and warm water to remove

Therapies for chemical or biological toxins have improved, but getting these treatments to the people who need them can be difficult. The medical providers here are traveling through the Republic of Mali in western Africa to help the Red Cross of Canada give vaccinations to the populace. *(Nathalia Guerrero)*

the chemical from the skin and hair. Severe exposure usually requires additional decontamination of the gastrointestinal tract with laxatives. If medical care takes place quickly after exposure, the exposed individual should have a full and fast recovery. Once pesticides have been taken into the body, however, they resist most treatments designed to get them out of tissues.

Medical providers call upon a short list of drug therapies to work against pesticide poisoning. The anticonvulsive drug pentobarbital has been successful against certain insecticides called pyrethroid insecticides because it stabilizes membranes and their ion channels. Therapy for phosphate-containing organic insecticides includes the drug atropine, which restores the balance between acetylcholine and the enzyme cholinesterase, the two compounds responsible for communication between nerve cells.

Phosphate pesticides and chlorine pesticides present challenges in treatment. Phosphate-containing pesticides disrupt the normal functioning of nerve cell communication and can eventually cause permanent damage to the nervous system. Chlorine-containing pesticides cause a variety of cell damages and they do not respond to treatments. People exposed to either of these pesticide types should be maintain good respiratory activity by moving to a clean place and perhaps taking an anticonvulsive drug to treat seizures if they occur. All of these decisions are made under a doctor's care.

Industrial chemicals have a variety of chemical structures, and each group requires different treatments. Appendix E provides a brief overview of the types of therapies available for chemical exposures based on the manner in which these chemicals harm living tissue. As with all other environmental toxins, immediate treatment offers the best hope for recovery. Anyone exposed to toxins should be treated on-site as quickly as possible without wasting travel time to a hospital (until the victim is stabilized). If fast treatment is impossible, the person should be taken to a poison control center immediately.

Metal toxicity treatment focuses on removing as much of the persistent metal from the body as possible. Compounds called *chelating agents* capture metals by binding to the metal, as a claw picks up a stone. Chelating agents have been especially useful in treating lead poisoning. Two main chelaters used for this purpose are ethylenediaminetetraacetic acid (EDTA) and dimercaptosuccinic acid (succimer). Other chelating agents helpful in treating metal toxicities are dimercaprol, D-penicillamine, and edetate calcium disodium.

Extremely toxic nerve agents act on people and animals within minutes and so therapies are very limited. Any exposure to the following nerve agents requires immediate respiratory aid and an antidote: sarin, cyclosarin, soman, tabun, and VX. Treatments for nerve agents have not changed since 1945 when they were first created for possible use in World War II. The main treatment involves administering three drugs to counteract the effects of the nerve agent on the nervous system. The first drug is atropine, which belongs to a group of drugs called anticholinergic drugs and helps return normal function in nerves. The second drug, 2-pralidoxime chloride, reactivates normal nerve signal transfer and at the same time breaks the nerve agent into harmless compounds. Third, benzodiazepine drugs control seizures.

Of all the environmental toxins, people are least likely to confront lethal nerve agents, but they will likely be exposed during their lifetimes to pesticides, industrial chemicals, and solvents that have contaminated

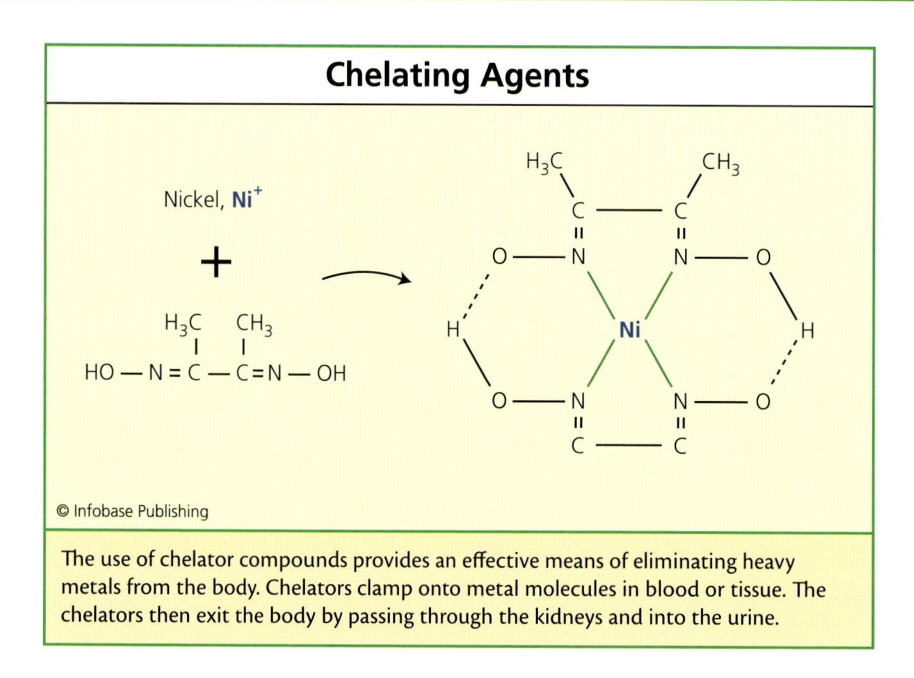

The use of chelator compounds provides an effective means of eliminating heavy metals from the body. Chelators clamp onto metal molecules in blood or tissue. The chelators then exit the body by passing through the kidneys and into the urine.

water, soil, or air. Fortunately, with a combination of medical care and awareness of the chemicals in the environment, most people can and do prevent serious illness caused by environmental toxins.

TOXIN POISONING PREVENTION

The most effective prevention for any toxin exposure is to avoid places where toxins are known to exist in high levels. Therefore, people should avoid known hazardous waste dump sites, industrial areas, polluted waters, pesticide spraying areas, and mines, and may even consider avoiding jobs in industries known to create toxic substances.

Most people have no way of knowing the amount of toxins they take into their bodies in a typical year. Preventive measures therefore have been designed to reduce exposure to high doses even if a person would be more likely to receive an infrequent low dose. People who work with toxins should use protective clothing, face masks or respirators, and gloves, use solvent-resistant creams on the skin, and try to confine their work to well-ventilated areas. Less toxic chemicals may be substituted for toxic chemicals in certain circumstances; for example, a weak acid substituting for a strong acid. These precautions are especially important for workers in the chemical industries or agriculture that uses pesticides.

People who do not work with chemicals have a much lower chance of exposure to high levels of hazardous substances. Most of the exposure that people receive likely comes from the air—indoor and outdoor—and water. People who are concerned about inhaling airborne chemicals should pay special attention to community alerts on air quality and stay in well-ventilated places. Residents of cities with very high levels of air pollution, such as Beijing, China, wear disposable masks to prevent inhalation of particles. These masks probably have limited usefulness against gases or very fine particles of less than 1 micrometer (μm).

Most municipal water systems in the United States supply a clean source of drinking water, but people with a high-risk health condition may want to take special precautions to avoid water pollutants. The easiest way to clean organic compounds out of tap water is to use a faucet filter or a portable filter, each of which removes many organic compounds when the water passes through the filter's carbon bed. Many people also purchase bottled water to replace tap water. In recent years, however, scientists have found bisphenol A and other compounds used in making plastics in bottled water and other plastic-bottled beverages. Pathology researcher Jack Bend of the University of Western Ontario in Canada told the *New York Times* in 2008, "The first thing is that [bisphenol A is] an endocrine disrupter, there's no question about that. Should people that are exposed to these low levels of this chemical be outrageously concerned? . . . We simply don't know. But we should find out." Though biomonitoring studies show that many, if not most, people have bisphenol A in their bodies, no solid evidence has emerged to prove that this is a health threat.

Home gardeners who use chemical pesticides may unleash another unwanted and preventable source of toxin exposure. Many people use pesticide sprayers recklessly, manner, wasting the product and putting more pesticide into the environment than they should. They should instead use efficient applicators that produce only the amount of pesticide needed and resist leaks. Home gardeners as well as commercial nurseries can also employ *biopesticides* in place of chemical pesticides. Biopesticides originate in the environment and are made by bacteria, fungi, or insects, and since biopesticides are natural substances, nature quickly degrades them. Some biopesticides are a bit less efficient at killing pests than chemical pesticides, but they offer significant advantages to human and environmental health.

CONCLUSION

Toxic chemicals in the environment worry many people because of their potential health threats to living things and the environment. Environmental medicine emphasizes the search for better ways to diagnose chemical toxicity and treating it, but new toxic chemicals seem to accumulate on Earth faster than science can detect them. Environmental medicine has been split in recent years over the best approach to solving the daunting problem of environmental toxins. Environmental medicine specialists must overcome challenges presented by today's toxins: chemicals that may come from either point sources or nonpoint sources, cause acute exposure or chronic exposure, or cause different toxicity symptoms. Because of these characteristics, the origin point of environmental illnesses is difficult to locate, and the illnesses difficult to diagnose. Fortunately, the equipment used in science today for detecting chemicals in the environment can find very small amounts of chemicals amid other materials. Diagnostic methods also improve almost daily.

Not all toxins possess the same potential to damage living things, so environmental medicine classifies these substances into groups based on their hazards to people and the environment. Most concerns regarding environmental toxins center on pesticides, organic solvents, and heavy metals. Newer research may soon add plastics and their components to this list of unhealthy substances. These compounds may soon be confirmed as endocrine disrupters, a new concern in environmental health and one that requires much more study. For the present, people know that the environment holds things that are probably not good for them, but the lack of convincing data leaves many discussions on these chemicals full of unanswered questions.

People can greatly reduce the threats that come from hazardous chemicals. First, the body repels many injuries from chemical toxins if people stay healthy and avoid known health threats. Second, scientists gain more knowledge monthly on environmental toxins and governments make slow but steady progress in putting controls on the most dangerous chemicals. People can do their part by learning about the potential toxins in their environment, followed by measures to prevent exposure. Third, the public can encourage the work of environmental organizations that focus on environmental health threats. Such organizations serve as an important resource for information, as a communicator with government, and as an activist in preventing further industrial wastes from entering the environment.

At home people can take steps to reduce their exposure to toxins. These actions include avoiding heavy use of pesticides, avoiding contaminated foods and drinks, avoiding direct contact with toxins, and perhaps finding a job that reduces the chances of exposure. Whatever approach people choose, each individual must decide how much toxin they are willing to accept in their lifestyle. With good information and sound decisions, most people can prevent serious harm from environmental toxins.

Hazards in the Air

The Earth's atmosphere contains several layers but the two layers closest to the Earth's surface, the troposphere and the stratosphere, determine the quality of the air used by living things. The troposphere extends from the Earth's surface to about 10 miles (16 km) above sea level. The stratosphere lies above the troposphere and extends to 30 miles (48 km) above sea level. An area high in ozone gas (O_3) lies where the troposphere and stratosphere meet. This ozone layer prevents living things on Earth from receiving too high an exposure to the Sun's ultraviolet radiation, which damages deoxyribonucleic acid (DNA) and causes some cancers. The combination of the ozone layer and the troposphere determine the quality of the air that people and animal life require.

The air used by Earth's living things contains 78.08 percent nitrogen, 20.95 percent oxygen, and 0.9 percent argon, plus other gases called *trace gases* because (aside from water vapor) they are found in low concentration in the atmosphere—0.038 percent or less. Trace gases consist of ozone, carbon dioxide (CO_2), water vapor (H_2O), methane (CH_4), nitrous oxide (N_2O), and *halons* consisting of chlorofluorocarbons (CFCs) and bromine-containing CFCs. Nitrous oxide and nitrogen dioxide belong to a group of compounds known as nitrogen oxides. Carbon dioxide, water vapor (1 percent of the atmosphere), methane, CFCs, and nitrous oxide in addition to nitrogen dioxide (NO_2) and fluorinated gases (such as hydrofluorocarbons) collectively make up greenhouse gases, which are components of the atmosphere that hold in heat and keep the troposphere warm.

Greenhouse gases contribute to global warming by preventing the Sun's heat from escaping back into space. Greenhouse gases, however, make up a small percentage of the composition of air and they present only part of the threats to air quality. The air also accumulates particles, soot, smoke, metals, allergy-causing material, photochemicals, ozone, and sulfur emissions from burning fossil fuels. Photochemicals are compounds produced when certain gases react in the atmosphere with sunlight.

Three main types of air pollution affect human and wildlife health: particle pollution, noxious gas pollution, and indoor air pollution. This chapter reviews each of these types of air pollution and the specific health threats they cause, the current trends in the Earth's air quality, and the technology and policies for improving the quality of the atmosphere. This chapter includes additional topics that have grown into health issues within the past few decades: noise pollution, electromagnetic fields, thermal air pollution, and allergies.

Strict air pollution laws have lowered emissions in the United States, but other parts of the world continue to produce heavy air pollution. Large cities in China have been plagued by continual air pollution hazards. Much of this pollution has come from China's rapid industrial growth and a dramatic increase in the number of coal-fired power plants and of vehicles. *(photoeveywhere.co.uk)*

EARTH'S AIR QUALITY

Air quality varies across the globe and tends to become poorer in industrialized regions. For this reason air pollution causes a special concern to people living in central Europe, eastern Asia, and northeastern North America where there are dense areas of industrialization. In the United States, the Clean Air Act of 1990 directs the U.S. Environmental Protection Agency (EPA) to set limits for pollutants in the air and to enforce those limits, also called air pollution *standards*. Hundreds or even thousands of pollutants may be in the air at any one moment, so the EPA measures six pollutants to represent total air quality: particulate matter (soot and dust), carbon monoxide (CO), sulfur dioxide (SO_2), nitrogen oxides, lead, and ozone.

Because the EPA has responsibility for controlling almost 200 different air pollutants, it sets strict standards for these six pollutants that will provide a picture of total air quality. These six pollutants, called *criteria pollutants,* are summarized in the following table. The EPA calls them criteria pollutants because their limits have been set according to criteria based on health and environmental studies. Therefore, air pollution experts know more about criteria pollutants than they do about many of the other compounds that pollute the air.

The six pollutants monitored by the EPA also occur in other parts of the world. Asia has developed air pollution that may be a serious health threat. In 2008 the National Institutes of Environmental Health Sciences presented findings from a study in Bangkok, Thailand, in which mortality rates correlated to pollution levels. Of course, air pollution does not stay in one place; it moves to other places in a process called *transboundary pollution.* University of California–Davis atmospheric scientist Steven Cliff told the *Los Angeles Times* in 2007, "The air above Los Angeles is primarily from Asia. Presumably that air has Asian pollution incorporated into it. More stuff starting up over there means more stuff ending up over here." The major air pollutants found in the United States and other countries are shown in the table on page 95.

The air contains two types of pollution based on source: primary air pollutants and secondary air pollutants. Primary air pollutants are substances emitted directly into the air, such as carbon monoxide from vehicles. Secondary air pollutants form in the atmosphere when two or more primary pollutants react. Acid rain and ozone are secondary air pollutants.

The EPA's Criteria Pollutants for Assessing Air Quality			
Pollutant	**Description**	**Health Effects**	**Main Sources**
$PM_{2.5}$ particulate matter	solid particles measuring about 0.0001 inches (2.5 μm) and smaller	lung diseases (asthma, chronic bronchitis, respiratory infections) heart disease (heart attack, irregular heartbeat)	fires, road dust, electricity generation (from burning fossil fuels), industry
PM_{10} particulate matter	solid particles measuring between 0.0001 inches (2.5 μm) and 0.0004 inches (10 μm)		road dust, other dust sources, fires
carbon monoxide	odorless, colorless gas	forms carboxyhemoglobin in the blood that prevents cells from getting oxygen	inefficient combustion engines in vehicles
sulfur dioxide	metal-containing gas released by burning coal or petroleum fuels	asthma, heart, and lung disease in the elderly, respiratory illnesses in children	electricity generation, fossil fuel combustion
nitrogen oxides	highly reactive gases: nitrous oxide, nitrogen dioxide, nitric oxide (NO), nitric acid (HNO_3), nitrate compounds (X- NO_3)	respiratory problems, reacts to form other toxic compounds in the air	vehicles, electricity generation, machinery, fossil fuel combustion, industry
lead	particles containing this heavy metal found naturally in the earth	damage to brain, nerves, and the cardiovascular, reproductive, and muscular systems	manufacturing plants, non-road equipment, electricity generation, fossil fuel combustion
ground-level ozone	naturally occurring gas made of three oxygen molecules; dangerous when at ground level rather than high altitude	respiratory irritation	vehicles, industry, gasoline vapors, chemical solvents

In addition to Asia, Europe has fought a continuing battle against air pollution since industries began growing there in the 1930s. In 2002 the European Union issued *The Sixth Environment Action Program of the European Community 2002–2012* to outline objectives for cleaning up Europe's air pollution to the year 2012. The objectives are as follows:

- carbon dioxide concentration below 550 ppm
- reduction of greenhouse gases by 70 percent compared with 1990 levels
- reduction of greenhouse gas emissions from transportation, including aircraft and ships
- prevention and reduction of methane emissions
- reduction of greenhouse gases from vehicles, with an increase in the use of alternative fuels
- phasing out the use of industrial fluorinated gases including hydrofluorocarbons (HFCs)
- setting and achieving goals in reducing ozone, particles, and indoor air pollution

World carbon dioxide emissions have become a benchmark for all global air pollution because they correlate to the presence of industrial emissions and a large number of vehicles. The following table illustrates the wide variation of carbon dioxide emissions in the world.

The world's total carbon dioxide emissions have been increasing for the past several decades. At present all regions of the world pour about 31,970 tons (29,000 metric tons) of carbon dioxide into the atmosphere each year.

MEASURING AND FORECASTING AIR QUALITY

Air quality monitoring has made use of global imaging to assess the density of pollution that covers continents or the entire globe. In global imaging, a satellite equipped with an instrument called a *spectrometer* detects different wavelengths of the Sun's light as it reacts with substances in the atmosphere. The European Space Agency (ESA) operates *Envisat,* a satellite carrying analytical instruments that monitors a 596-mile (960-

Major Air Pollutants	
Pollutant	**Main Sources**
particulate matter	industry, construction, incinerators, mining and quarries, metals processing, fuel combustion, forest fires
carbon monoxide	vehicles, heavy and small equipment, waste disposal facilities, metals processing, chemical industry
sulfur dioxide	electric utilities, industry, heavy equipment, petroleum refineries
hydrogen sulfide	oil wells, oil refineries
nitrogen oxides	vehicles, heavy equipment, electric utilities, industry
volatile organic compounds (VOCs)	solvent-using processes, vehicles, heavy equipment, transportation, fuel evaporation
lead particles	leaded-fuel combustion, smelters, battery plants
acid rain	sulfur dioxide and nitrogen oxides combines with compounds in atmosphere to create acids: sulfuric and nitric, respectively
ozone	reaction of nitrous oxide with VOCs; also vehicles, factories, landfills

km) band of the Earth as it covers the entire planet every six days. Sunlight may do one of three things when it hits a compound: (1) reflect in the opposite direction; (2) pass through the substance in a process called *transmittance*; or (3) scatter in many directions. Each of these distortions to sunlight's path changes the light's wavelength, and each wavelength of the light spectrum correlates with a different color. The spectrometer in a satellite detects and measures the various wavelengths in the ultraviolet, visible, and infrared portions of the spectrum, and the instrument then interprets the wavelengths as colors. A computer generates a map containing the colors indicating regions of good to bad air quality.

EXAMPLES OF WORLD CARBON DIOXIDE EMISSIONS	
COUNTRIES OR REGIONS	MILLION TONS (METRIC TONS)
Afghanistan	1.1 (1.0)
Antarctica	0.28 (0.25)
Brazil	397 (360)
Canada	696 (632)
China	5,870 (5,325)
Denmark	56 (51)
Germany	931 (845)
India	1,290 (1,170)
Russia	1,874 (1,700)
Saudi Arabia	454 (412)
Somalia	0.83 (0.75)
United Kingdom	639 (580)
United States	6,570 (5,960)
Source: U.S. Department of Energy	

For the United States, National Aeronautics and Space Administration (NASA) satellites measure wavelengths correlated to single pollutants. For example, a satellite's spectrometer differentiates carbon monoxide levels: Red indicates high concentrations of 450 parts per billion (ppb); blue indicates low carbon monoxide levels of 50 ppb. The colors between these two ends of the spectrum represent intermediate carbon monoxide levels. NASA scientists then use atmospheric wind speeds to create a video of pollution as it moves over the land and oceans. Atmospheric scientist John Gille explained, "With these new observations you clearly see that air pollution is much more than a local problem. It's a global issue. Much of the air pollution that humans generate comes from natural sources such as large fires that travel great distances and affects areas far from the source." Global assessments such as those by the ESA and NASA have helped environmental scientists build a more complete picture of air pollution than ever before.

People depend on air pollution information on a more daily basis than satellite data by referring to their local *Air Quality Index* (AQI) produced by the EPA. Other countries have similar indices. The EPA's system measures five air components to calculate AQI: ozone, carbon monoxide, sulfur dioxide, $PM_{2.5}$, and $PM_{1.0}$ (particles of about 1 μm). Local health officials then convert these measurements into a numerical score to create the AQI and give each category a standard color code to make the scores easy to interpret by the public. The following table provides a description of the AQI.

The Air Quality Index			
AQI	Health Concern Level	Health Concern Description	AQI Color Code
0–50	good	little or no risk	green
51–100	moderate	slight health concern for some people	yellow
101–150	unhealthy for sensitive groups	health effects in high-risk health individuals	orange
151–200	unhealthy	everyone experiences a health effect	red
201–300	very unhealthy	health alert; possibly serious	purple
302–500	hazardous	serious health alert	maroon

Ozone and ground-level particles present the biggest health hazards on the AQI. Ozone in the stratosphere protects people and animals from ultraviolet radiation, but air pollution has caused much of this ozone to become trapped at lower levels in the troposphere so it cannot reach the stratosphere. People inhale this ozone, which causes coughing, irritation, and lung damage. Inhaled ozone also worsens bronchitis, emphysema, and asthma in people with these illnesses, and long-term exposure may scar the lungs. Because ozone has both good effects and harmful effects in the atmosphere, the EPA uses the catchphrase "Good Up High, Bad Nearby" to describe ozone.

Local ozone levels can be measured using data-collecting aircraft, balloons, or portable devices used on the ground. Ozone-measuring balloons carry an instrument called an *ozonesonde,* which takes in a set volume of air and measures ozone by either detecting light wavelengths or by chemical reactions. Handheld devices contain either a spectrometer or a sensor that detects ozone by measuring the ability of the air sample to conduct an electrical charge—greater conduction equals higher levels of ozone.

Air pollution has caused visible changes in the environment. The speckled light gray peppered moth has evolved from the light speckling shown here to a much darker version since the Industrial Revolution. Dark moths possess coloring that blends in with surfaces dirtied by soot and smog. In some industrialized places, 90 percent of the moths are dark, which was once a very rare mutation.

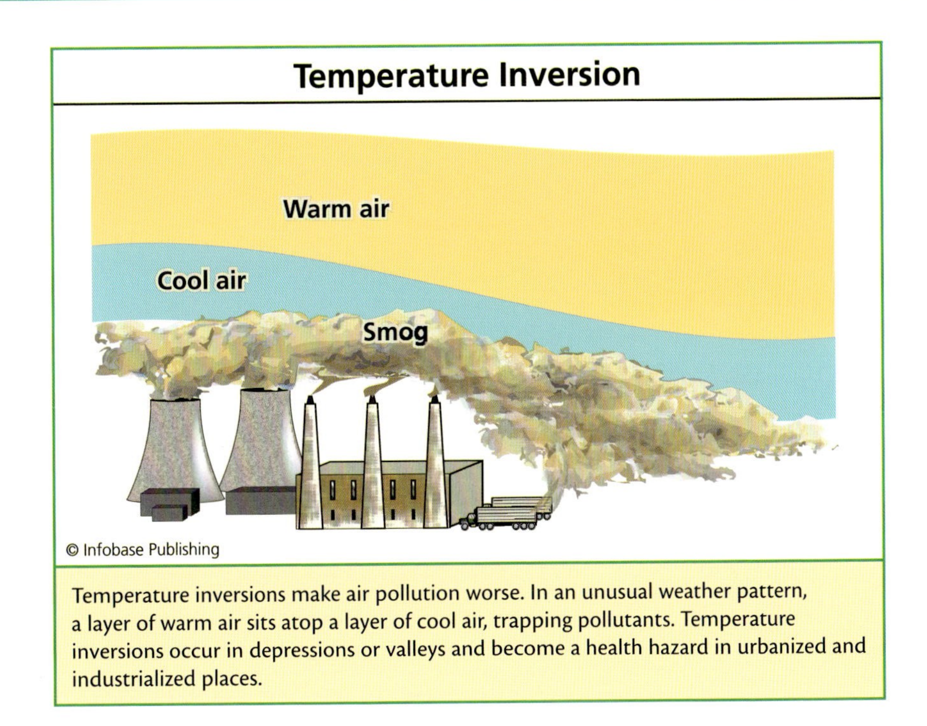

Temperature inversions make air pollution worse. In an unusual weather pattern, a layer of warm air sits atop a layer of cool air, trapping pollutants. Temperature inversions occur in depressions or valleys and become a health hazard in urbanized and industrialized places.

The EPA categorizes air particles into two groups according to size: (1) inhalable coarse particles of 2.5–10 μm in diameter, which are dense near roadways and construction sites, and (2) fine particles of 2.5 μm in diameter or smaller, emitted from forest fires, power plants, manufacturing plants, and cars as exhaust, gas emissions, smoke, or haze. Scientists monitor the amount of total particles in the air with sensors that contain a laser light source. As the instrument takes in a volume of particle-laden air, the air moves through the sensor and particles intercept the laser beam. Each disruption in the beam equals one particle—the instrument reports the results as particles per volume.

Temperature inversions affect the amount of air pollution. Warm air at the Earth's surface usually rises and mixes with the cool air at higher altitudes. This mixing creates local winds and air turbulence. On occasion the warm layer does not mix, but merely sits atop a cool air layer nearer the ground. This situation is called an inversion because the air becomes warmer as the altitude rises, which is the opposite of the normal circumstance of cooler temperatures at higher altitudes. The dense, cool air traps pollutants that would normally flow upward and disperse. As a result the pollution builds up in the air, especially over metropolitan areas located in valleys, which further traps the cool air. Temperature inversions increase the concentration of air pollutants and aggravate lung ailments such as asthma.

INDOOR AIR QUALITY

People in industrialized nations spend at least 65 percent of their time indoors and another 25 percent in indoor environments such as cars, buses, or airplanes. Indoor air can be more polluted than outdoor air, even in communities that have little or no industry. This indoor air pollution comes from two sources. First, particles, dust, pollen, pesticides, and gases from the outdoors enter through open windows and doors. The second source of indoor air pollution comes from diverse substances emitted from things found inside homes and offices: carpets, furniture, heaters, gas stoves, fireplaces, cleaning products, paint, wall coverings, indoor pesticides, improperly stored chemicals, and tobacco smoke. Some homes may also contain radon and carbon monoxide. Radon is a naturally occurring radioactive gas that can cause cancer, and carbon monoxide is an odorless gas that leaks from faulty gas home heaters, stoves, kerosene space heaters, or cars idling in a closed garage.

Indoor air pollutants are associated with health problems, called sick building syndrome, in people and pets. These pollutants include the following:

- radon—from rocks underground or from building materials
- molds—from wet or moist items, the outdoors, or dirt carried on shoes, clothing, or pets
- pollen—from outdoor air, indoor plants, or carried on clothing or pets
- carbon monoxide—possible hazard from faulty heaters or other items
- nitrogen dioxide—from kerosene heaters, gas stoves, and tobacco smoke
- formaldehyde and other organic compound vapors—from furniture, paints, wood preservatives, disinfectants, automotive products, and dry-cleaned clothing
- pesticides—from garden, cropland, lawn, produce, and indoor pesticide products
- lead—from lead-based paint and contaminated soil and drinking water

When a local AQI is high, news outlets warn that people with respiratory problems should avoid spending time outside. Unfortunately, many buildings contain air of a quality equal to or worse than heavily polluted outdoor air. This predicament is discussed in the "Indoor Air Quality" sidebar.

- asbestos—possibly in deteriorated older buildings or damaged insulation or fireproofing
- tobacco smoke—possible in buildings that admit smokers or from outdoors

Richard Corsi, an expert on indoor contaminants at the University of Texas, discussed the unique problem of indoor air pollution with the *San Francisco Chronicle* in 2004, saying that "We spend so much time indoors, yet spend so little time thinking about the quality of our indoor air." But people can use at least three tactics to improve their indoor air quality: (1) source control; (2) better ventilation; and (3) air cleaners.

Source control involves measuring for radon and carbon monoxide and eliminating their sources, if possible. Builders should select carpeting, floor coverings, cabinetry, furniture, and wall coverings that emit no harmful compounds. People can also limit the amount of chemical sprays used indoors, and store chemical products in a single location far from household members.

Proper ventilation consists of opening windows and doors, using a window air conditioner that vents outside, installing fans in the bathroom and kitchen that exhaust outdoors, installing air-to-air heat exchangers, and moving as many pollution-causing activities outdoors as possible. Activities such as painting, varnishing, gluing, using kerosene appliances, changing vacuum bags, welding, soldering, or sanding should be done outdoors when possible.

Air cleaning devices collect pollutants from the air and trap them on a filter, which the owner simply changes when full. These precautions are especially important in newer buildings in which strong window and door seals create an airtight environment.

Indoor air pollution in developing countries arises from a different set of circumstances than found in the industrialized world. In these places, indoor cooking often uses wood, crop wastes, dung, or coal, which contributes chemicals and particles to the air. The World Health Organization has estimated that indoor air pollution kills 1.6 million people yearly worldwide—one death every 20 seconds—with women and children who spend the most time indoors at highest risk. Many people may overlook indoor air when they think of air pollution, but this is an important concern in environmental medicine.

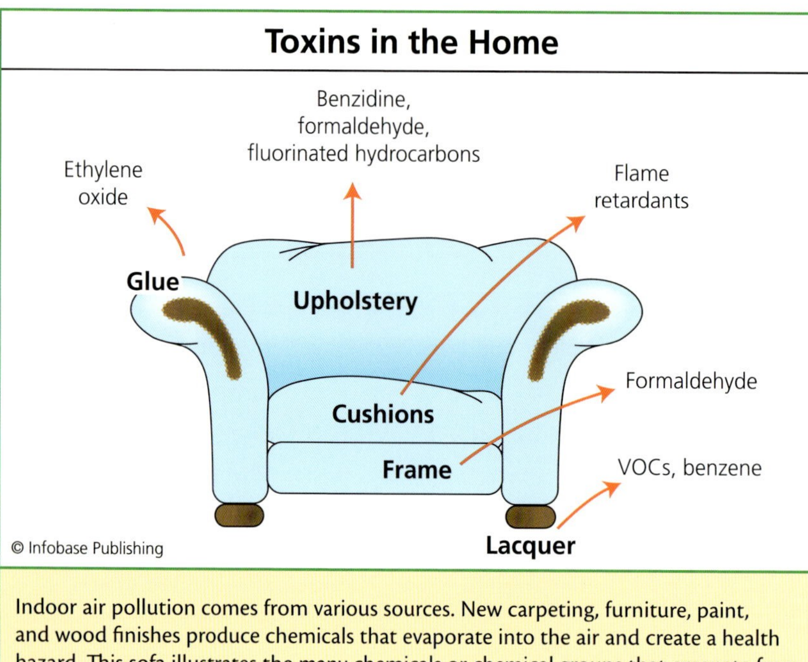

Indoor air pollution comes from various sources. New carpeting, furniture, paint, and wood finishes produce chemicals that evaporate into the air and create a health hazard. This sofa illustrates the many chemicals or chemical groups that emanate from it for many months.

GREENHOUSE GASES AND THE OZONE LAYER

The greenhouse effect and the stratosphere ozone layer represent two components of the Earth's atmosphere that sustain life. Oxygen, carbon dioxide, water and *nutrient cycles*—called biogeochemical cycles—are the other components that biota require for life.

The greenhouse effect is a natural occurrence in which certain gases in the atmosphere regulate the temperature of the Earth's surface. Greenhouse gases and water vapor allow the Sun's energy to pass through the atmosphere and reach the planet's surface, but they retain a portion of the heat that reflects back from the surface in the form of long-wavelength light. The gases then release heat energy from the light they absorb and the reflected light. The effect has been named the greenhouse effect because the absorption, reflection, and heating resemble the actions that take place inside plant greenhouses. The resulting heat in the atmosphere sets up a comfortable temperature range that supports the evolution of life on Earth;

without the greenhouse effect, the Earth's average temperature would be about - 0.4°F (-18°C) and the oceans would freeze.

Ozone also plays a protective role by absorbing the Sun's ultraviolet-B (UVB) radiation, thereby shielding biota from its DNA-damaging effects. About 10 percent of atmospheric ozone lies in the troposphere, much of it formed when nitrogen oxide emissions in car exhaust react with sunlight to form nitrogen dioxide and oxygen, as follows:

$$NO_X + O_3 \rightarrow NO_2 + O_2$$

This tropospheric reaction causes *ozone smog,* which does not protect people from UVB but instead irritates the eyes, nasal passages, and throat. Ozone smog can also reduce resistance to colds and may aggravate chronic diseases such as asthma, bronchitis, and heart disease.

While ozone smog exerts these unhealthy effects, other emissions destroy some of the beneficial ozone in the upper atmosphere: chlorofluorocarbons (CFCs) and halon flame retardants such as carbon tetrachloride, methyl bromide, and methyl chloroform. In the 1970s scientists had developed instruments to study the Earth's ozone layer. During the 1980s, they began noticing a marked thinning of the ozone layer and a hole in the ozone above Antarctica began enlarging. By 1991 the hole covered the entire Antarctic continent.

Beginning in 1986, several national governments banned the use of CFCs for the purpose of halting further damage to the Earth's protective ozone layer, but their effects would not be instantaneous. Ozone weakening continued until the end of 2005 when the ban finally seemed to be making a positive effect; the Antarctic ozone hole stopped growing. David Hoffman of the U.S. National Oceanographic and Atmospheric Administration (NOAA) told *BBC News* in 2006, "I'm very optimistic that we will have a normal ozone layer sometime, not in my lifetime, but perhaps in yours." Though the layer no longer decreases near the South Pole, neither is it improving. Ozone holes now appear each summer over western Siberia's Ural Mountains and over an area covering the Baltic nations (Latvia, Lithuania, and Estonia), northeast to Moscow, and northwest to St. Petersburg. Saving beneficial ozone continues to be a long-term goal.

Though the greenhouse effect helps life on Earth, too much warming has led to the unhealthy condition now recognized as global warming. Until humans arrived on the planet, average amounts of atmospheric carbon dioxide fluctuated between 200 ppm and 280 ppm for 650,000 years. Carbon dioxide levels have risen 40 percent to 387 ppm since the Industrial Revolution and they may exceed 600 ppm in less than 50 years at their present rate of production. The rise in the Earth's average temperature has correlated with this rise in carbon dioxide levels. Carbon dioxide receives the most blame for global warming, but other substances contribute as much or more to global warming. The table on page 105 lists today's most troublesome greenhouse gases and their effect on global temperatures, described as Global Warming Potential (GWP). GWP equals the global warming caused by 2.2 pounds (1 kg) of a gas relative to the same amount of carbon dioxide. For example, in 20 years methane raises the troposphere's average temperature 62 times more than the same amount of carbon dioxide.

In 2008 Martin Perry of the Intergovernmental Panel on Climate Change warned the United Kingdom's *Guardian,* "Despite all the talk, the situation is getting worse. Levels of greenhouse gases continue to rise in the atmosphere and the rate of that rise is accelerating. We are already seeing the impacts of climate change and the scale of those impacts will also accelerate, until we decide to do something about it." The Kyoto Protocol, highlighted in the sidebar on page 106, represented one important step to do something about climate change.

CARBON DIOXIDE

Carbon dioxide is a colorless, odorless greenhouse gas composed of one carbon molecule and two oxygen molecules. Carbon dioxide makes up only 0.04 percent of the atmosphere but it is essential for photosynthesis performed by plants, algae, and some bacteria. Photosynthesis converts carbon dioxide to oxygen and it is the main source of oxygen in the atmosphere. Natural levels of carbon dioxide in the atmosphere contribute to the use and reuse of carbon on Earth, known as the *carbon cycle.* Carbon is a central component in all living things on Earth, and the carbon cycle helps conserve this element.

The atmosphere's current carbon dioxide level of 387 ppm is the highest concentration this gas has reached in 160,000 years. If carbon dioxide levels reach 550 ppm within the next 50 years, as predicted, the world's average surface temperature will increase by 1.8 to 6.3°F (1–3.5°C). The world's normal temperatures fall into a very wide range from the warm equator to cold polar regions. A rise of only one or two degrees in the planet's average

The Major Greenhouse Gases and Their Effect on Global Warming

Greenhouse Gas	Time in the Troposphere before Degrading (years)	Global Warming Potential	
		20 years	*100 years*
carbon dioxide	100	1	1
methane	12	62	23
nitrous oxide	115	275	296
chlorofluorocarbons (CFCs)	55–550	5,500	5,800
hydrochlorofluorocarbons (CH_xF_x)	1–260	100–9,400	100–12,000
hydrofluorocarbons (C_xF_x)	3,000–50,000	6,000–15,000	8,600–22,000
ethers (CH_3OCH_3)	0.015	1	1
halogenated ethers (CH_xOCF_x)	2.5–100	100–13,000	30–15,000

Source: University of Michigan Global Change Program

temperature therefore signifies a dramatic increase in global temperature. This global warming effect is demonstrated by the following estimates:

- At 1.8°F (1°C) global temperature increase, animals begin migrating toward the poles and high-elevation forests begin to disappear.
- By a 3.6°F (2°C) increase, 30 percent of species are threatened.
- By a 5.4°F (3°C) increase, 30 percent of all wetlands disappear.
- By a 7.2°F (4°C) increase, 40 percent of all species are extinct.

Only drastic and immediate cuts in carbon dioxide emissions can reverse this trend.

THE KYOTO PROTOCOL

In 1997 representatives from 161 nations met in Kyoto, Japan, to sign a treaty for the purpose of controlling global warming. This treaty, called the Kyoto Protocol, puts forth a plan to address global warming by making the following recommendations:

- About 40 industrialized countries were to cut their carbon dioxide, methane, and nitrous oxide levels in the air by 5.2 percent below 1990 levels.
- The target date for the developed countries' objectives was the year 2012.
- Poorer developing nations were not required to lower their greenhouse gas emissions until later.
- Participating countries could conduct greenhouse gas trading.
- Governments could issue fines to businesses that exceeded their upper limit of emissions.

In 2004 member nations ratified the Kyoto Protocol to address practical problems of achieving national goals and the treaty went into effect on February 16, 2005. (Ratification refers to a formal signing, approval, and agreement to abide by the treaty's conditions.) Since the Kyoto Protocol was opened for signature in 1997, more than 180 countries have ratified it. The United States and Australia have resisted ratifying the protocol, citing a number of flaws in the agreement, particularly the provision for excluding China and India from the emissions limits even though they have two of the fastest-growing industrial economies in the

Carbon dioxide makes up about 85 percent of all emissions from human activities. The ranking of the top carbon dioxide–producing nations varies every few years, but the following countries consistently top the list: The United States, China, Russia, Japan, India, Germany, Canada, the United Kingdom, South Korea, and Mexico. Worldwide, fossil fuel combustion by cars, trucks, ships, boating, off-road vehicles, and other transportation serves as the largest single source of carbon dioxide, about 40 times greater than the next biggest carbon dioxide source, which is fossil fuel use in manufacturing processes. The following listing by the EPA ranks in order the major producers of carbon dioxide today: (1) fossil fuel combustion; (2) nonenergy fuel use; (3) iron and steel production; (4) cement manufacture; (5) natural gas systems; (6) municipal solid waste combustion;

world. Although the United States had remained part of the treaty discussions, it withdrew its participation completely in 2001, unhappy with what the George W. Bush administration felt was vague language on emissions control. Robert Donkers, an environment counselor for the European Union, told the *Washington Post* at the time of the U.S. withdrawal, "It is not just the European Union versus the United States. This is Australia and the United States against the rest of the world." Despite world criticism, the United States denounced the treaty because developed nations would bear too much of the burden in emissions reduction. White House spokesman Ari Fleischer explained President George W. Bush's position to the *New York Times*, "The president has been unequivocal. He does not support the Kyoto treaty. It is not in the United States' economic best interest." Since then developing nations such as Mexico and Brazil have taken much stronger steps than in the past toward stopping climate change. For this reason the United States may change its stance on global warming and the Kyoto Protocol in the near future.

Even with flaws that some leaders see in the Kyoto Protocol, this treaty represents a key step in international cooperation toward reducing global warming, though the treaty's success without the United States' participation remains in question. "The greatest value is symbolic," said global warming advocate Eileen Claussen about the treaty. Agreements such as the Kyoto Protocol cannot solve the global warming problem all by itself. This treaty may help, however, as one of many approaches to reducing greenhouse gases and putting a halt to global warming. A future U.S. administration may decide to sign the Kyoto Protocol to build international unity in the fight against global warming.

Places that were at one time pristine areas now contain pollution. Haze covering the Grand Canyon comes from power plant emissions, vehicle traffic, and west coast urban air pollution carried by winds. *(Kris Hanke, courtesy of iStockPhoto.com)*

(7) lime production; (8) ammonia production and urea consumption; (9) limestone and dolomite use; (10) croplands; (11) soda ash (sodium carbonate) manufacture; (12) aluminum production; and (13) petrochemical production.

Industries supply the world with the products that help people live or to simply make life enjoyable. But the health of the environment and the health of living things pay a price for conveniences manufactured by industries large and small. Robin Oakley, head of Greenpeace's climate change campaign, told the United Kingdom's *Guardian* in 2008, "We're now witnessing a key moment in the climate change story, and it's not good news. The last time the atmosphere was this choked with carbon dioxide humans were yet to evolve as a species." Carbon dioxide currently represents the single biggest focus area in environmental studies.

The task of reversing the upward trend of carbon dioxide in the atmosphere seems daunting if not impossible. Hopes rest on new technologies in transportation, alternative fuels, sustainable manufacturing, and the cooperation of industries and nations to make these things possible. Cutting-edge technologies might also soon emerge such as huge air scrubbers stationed across the landscape to pull excess carbon dioxide out of the air. Global Research Technologies has experimented with this idea in Tucson, Arizona. Company president Alan Wright gave a glimpse of the promise as well as the scale of atmosphere air scrubbers when he told *Time* magazine in 2008, "If we built one [a scrubber] the size of the Great Wall of China and it removed 100 percent of the carbon dioxide that went through it, it would capture half of all emissions in the world." It is encouraging to think that a single device could actually have such a dramatic effect on the environment, but the magnitude of such a project makes it very unrealistic. Global warming technology will likely need a combination of many different powerful techniques for controlling carbon dioxide and other greenhouse gases.

RADIATION

Radiation consists of energy in the form of either electromagnetic waves or particles that travel through the air. Electromagnetic waves occur when a magnetic field crosses perpendicularly with an electric field. This collision creates waves of different wavelengths—the distance between waves. Particle radiation occurs when an atom releases a subatomic particle: a positively charged proton (alpha particle), a neutral neutron, or a negatively charged electron (beta particle). (Gamma rays represent a fourth type of radiation.) Elements in the universe that emit any of these particles are said to be radioactive, and the term *radioactivity* refers to the disintegration of an element by emitting protons, neutrons, or electrons.

People exposed to increasingly large doses of radiation suffer radiation burns, radiation sickness, or cancer, described in the following table. Natural radioactivity comes from the elements radium, radon, and uranium, and human-created radioactivity comes from new elements created in nuclear reactors, such as plutonium. Radiation causes a variety of general health risks such as red blood cell damage, lymph system damage, injury to DNA, bone and bone marrow damage, hair loss, and skin damage. Many of these health effects occur only with exposure to high levels of radiation. Therefore, people living in areas with no known high-level sources have few health risks from radiation.

Radioactive substances of most concern in the environment are those associated with nuclear power generation—uranium and plutonium—and naturally occurring radium. Nuclear power generators use natural uranium in forms that produce new elements that are also radioactive. The nuclear reactors inside nuclear power plants can then be a source of radioactive uranium, plutonium, and hundreds of other by-products made during nuclear reactions to produce energy. These by-products contain dozens of *isotopes* of uranium and other elements, meaning their atoms contain an unusual number of neutrons. Well-managed nuclear power plants have safeguards to protect against the release of these elements into the environment, although accidents have occurred in the United States and abroad whereby radioactivity escaped into surrounding communities.

RADIATION AND ENVIRONMENTAL HEALTH		
RADIATION ILLNESS	**ABSORBED DOSE (GRAYS) ***	**HEALTH EFFECTS**
radiation burn	1–2	reddened skin, fatigue, nausea, headache
mild-moderate radiation sickness (or syndrome)	2–3.5	vomiting, diarrhea, bloody stool, hair loss, red blood cell damage
severe radiation sickness	3.5–8	nausea and vomiting within 30 minutes, disorientation, fever, low blood pressure, death
cancer	Any dose increases risk in later life	cancers in bone, lungs, and other organs, death

Source: Mayo Clinic

* A gray equals a quantity of absorbed radiation: 1 gray = 100 rad (radiation absorbed dose)

Uranium is the heaviest of the natural elements, a dense, silver-white metal found deposited in other minerals. In the United States, the Nuclear Regulatory Commission controls the use and sale of uranium for energy production and nuclear weapons. Earth's natural uranium consists mostly of uranium 238, which has a half-life of 4.6 billion years. (The isotope uranium 235 makes up a very small percentage of natural uranium.) This long half-life means that uranium possesses weak radioactivity because it emits alpha particles very slowly. As uranium decays it produces other radioactive elements, principally radon and plutonium.

People take small amounts of natural uranium into their bodies in food and water and by breathing it in. The body rids itself of most of this uranium, but a fraction remains in bone and stays there for many years. High levels of uranium cause more serious problems. It can damage the kidney as this organ tries to excrete it and can lead to bone and lung cancers.

The nuclear industry makes plutonium 239 from uranium 238 and uses this plutonium in fission reactions. Fission consists of the splitting of an atom's nucleus into two or more pieces with a simultaneous release of energy. Plutonium 239 has a half-life of 24,000 years and is toxic to humans and animals if it escapes from nuclear facilities into the environment. Inhaled plutonium moves from the lungs into the bloodstream and concentrates in bone and the liver and remains there for up to 50 years; the liver flushes plutonium out of its tissue slightly faster than bone can. High-level plutonium exposure leads to lung disease and cancers.

Radium is a rare, metallic element that has a half-life of 1,620 years, meaning it requires that many years for half of its radioactivity to disintegrate to safe levels. Radium forms radon gas as it decays, which presents a health concern because radon can seep into homes, buildings, and well water, and contributes to sick building syndrome.

Radon is a colorless, odorless, tasteless, and nonreactive gas. Radon diffuses through soil and drifts through the air; it commonly enters buildings by seeping through basement floors and walls, and stays there because it is heavier than air. Without an exchange of fresh air, radon builds up indoors and chronic exposure may lead to lung cancer. Radon produces

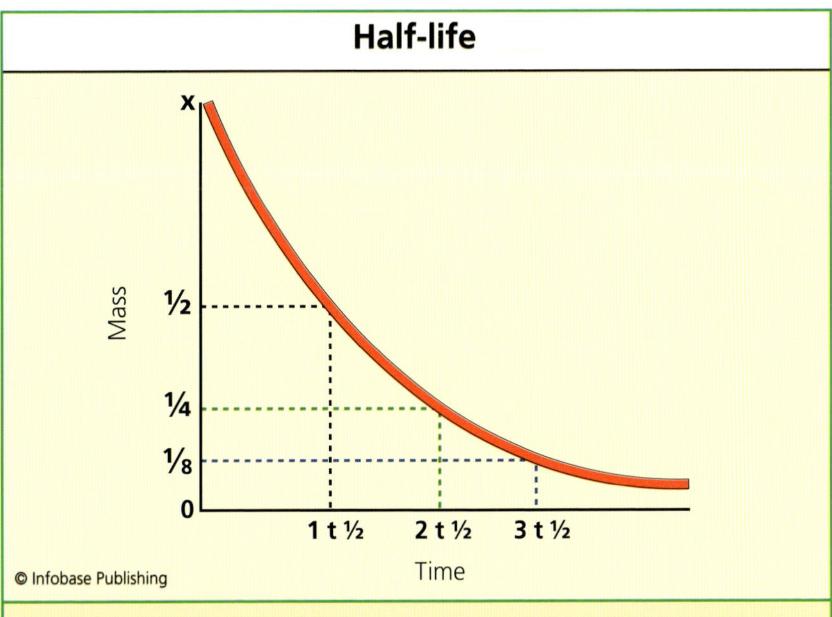

The half-life of a radioactive molecule is the time required for half of the atoms to decay to a more stable form. The Earth contains about 100 naturally radioactive elements. Krypton 90 possesses the shortest half-life at 33 seconds; rubidium 87 has the longest half-life at 47 billion years.

different *isotopes,* most of which have a half-life of no longer than four days. They decay into a variety of new isotopes that attach to dust and other particles in the air. One of these isotopes, lead 210, can be harmful when inhaled, and lead 210 furthermore has a half-life of 20.4 years. Lead 210 and other radon isotopes stick to the cells lining the lung's airways. In the lungs, the isotopes emit a constant stream of alpha particles, and this long-term exposure to radioactivity damages DNA. Radon may be a major cause of cancer worldwide. In the United States, radon is a major cause of lung cancer among nonsmokers.

ELECTROMAGNETIC FIELDS

The electromagnetic spectrum encompasses the following wave sources: electric fields, magnetic fields, alternating electric and magnetic fields, radio and television waves, microwaves, infrared light, visible light, ultraviolet light, and ionizing radiation or X-rays. Electric and magnetic fields occur at the low-frequency end of this spectrum and X-rays occur at the high-frequency end of the spectrum. Frequency is the number of waves generated per second. These forms of energy are usually described as fields because an area, or field, radiates outward from the source in all directions as opposed to moving in a straight line as an alpha particle does. An electric lamp, for example, generates a small electric field in the air around it due to the current flowing to the lamp; large electric generating stations create a similar but much larger electromagnetic field.

Growing demand for electricity and an increasing number of electric products have increased the potential number of electromagnetic fields people encounter each day. The human body itself contains electric current, such as the currents used to transport substances into and out of cells and the communication between nerve-cell endings. When weak electric charges in the body meet an electromagnetic field, the field alters the body's normal current. Small changes in the body's currents do not affect health, but higher energy fields can induce headaches, anxiety, depression, or fatigue.

The public's concern about high-voltage lines and health has grown since about 1980. Wisconsin's *Midwest Today* in 1996 quoted Robert Becker, physician and author of *Cross Currents,* a book that explores the dangers of electromagnetic fields: "Electromagnetic fields could turn out to be a far worse environmental disaster, affecting far more people, than toxic waste, radiation, or asbestos." This debate has not ended. While

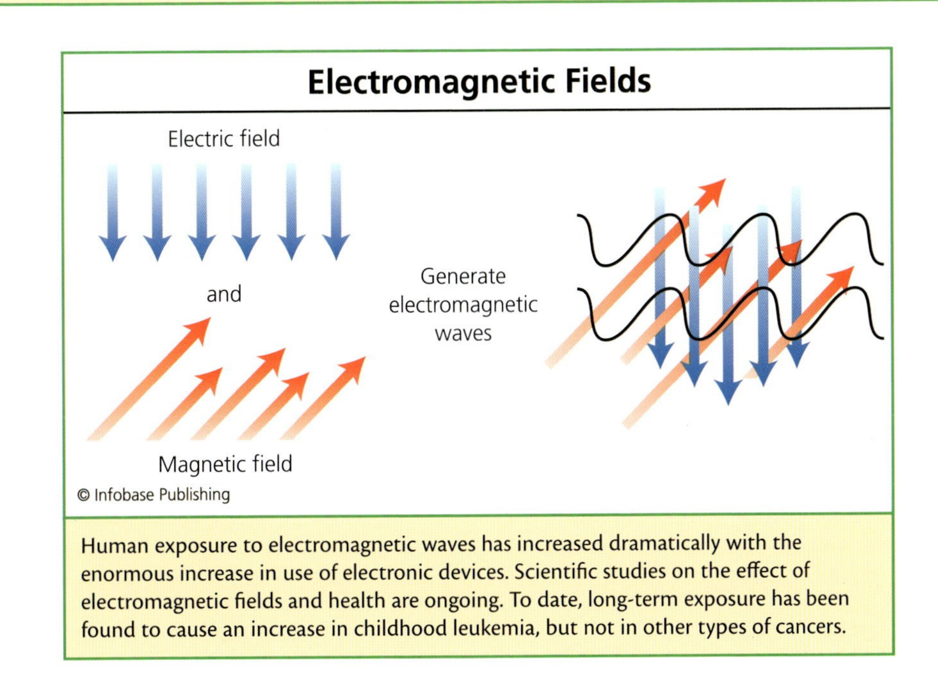

Human exposure to electromagnetic waves has increased dramatically with the enormous increase in use of electronic devices. Scientific studies on the effect of electromagnetic fields and health are ongoing. To date, long-term exposure has been found to cause an increase in childhood leukemia, but not in other types of cancers.

Medical News Today warned in 2004 that high-voltage lines may double the risk of leukemia in children living within 330 feet (100 m) of them, the Health Physics Society based in Virginia stated in 2008, "there are no known health risks that have been conclusively demonstrated in relation to living near high-voltage power lines." Medical researchers who have conducted studies in the effort to find a connection between power lines and disease generally agree with the Health Physics Society's statement. Scientists, nevertheless, measure the electromagnetic fields emitted by household appliances in order to add to their store of knowledge on high-power and low-power fields and human health. The table on page 114 lists electromagnetic fields emitted by various household appliances.

Electromagnetic field sources in addition to power lines are mobile phones, televisions, radios, electric trains and trams, security systems, and radar. Mobile phone radiation emissions are measured as Specific Absorption Rate (SAR), on which the United States and Europe have set upper limits at 0.2 to 1.6 watts per kg (W/kg) in 1.0 gram of body tissue (one kg equals 2.2 pounds). A person with greater body weight can safely absorb more radiation than a smaller person. Some of the digital phones in use today approach an SAR of 1.5 W/kg.

APPROXIMATE ELECTRIC AND MAGNETIC FIELD STRENGTH OF HOUSEHOLD APPLIANCES		
APPLIANCE	ELECTRIC FIELD STRENGTH (VOLTS/METER)[1]	MAGNETIC FIELD STRENGTH (μT)[2]
microwave oven	250	4–8 at 30 cm
refrigerator	120	0.5–1.0 at 30 cm
hair dryer	80	6–2,000 at 3 cm
coffeemaker	60	0.3–1.0 at 1 m
television	60	0.01–0.15 at 1 m
computer	14	less than 0.01 at 30 cm
dishwasher	12	0.6–3.0 at 30 cm
washing machine	10	0.15–3.0 at 30 cm
Maximum Recommended Limit	**5,000**	**100**

Source: The World Health Organization and the Consumer Law Page (URL: www.consumerlawpage.com)

[1] field strength is the intensity of an electric field measured in volts/meter (volts per 3.28 feet)

[2] a microtesla (μT) is a unit of measurement of a magnetic field; the Earth's natural magnetic field has a strength of 50 μT. 1 cm = 0.4 inches, 1 m = 3.28 feet

PARTICULATES

Air particles range from large visible matter to tiny invisible particles. A so-called large particle is considered to be any particle greater than 10 μm in diameter, or about one-seventh the width of a human hair. The EPA enforces limits only on particles in the air measuring less than 10 μm in diameter. The EPA also distinguishes between primary and secondary particles. Primary particles come directly from a point source such as a construction site, an agricultural field, a smokestack, or a forest fire, and these tend to be large particles. Secondary particles are very small particles that form in the atmosphere from chemical reactions or electrostatic attractions. These fine particles make up the majority of health risks from particulate air pollution.

Particles in the air affect both visibility and health. When sunlight hits tiny particles in the air, haze forms and the distance a person can see decreases. This decreased visibility is not a health hazard, but it indicates that potentially harmful particles are in the air. Haze-forming particles come from natural sources such as fires and windblown mater, or from human activities involving cars, burning fuel, electric power plants, and industry. Haze potentially injures the respiratory tract and may lead to a shortened life span. Respiratory ailments take the form of coughing, respiratory irritation, difficulty breathing, asthma, and bronchitis. Particles in the air also cause a higher chance of illness in people with existing heart ailments.

Pollen from plants makes up a unique category of airborne pollutants. Pollen consists of particles invisible to the unaided eye that drift on the air from plant to plant as part of plant reproduction. Each type of plant produces pollen with its own unique shape and these pollen shapes contain hundreds of sharp spikes that enable the pollen to stick to surfaces when it falls from the air. Of course, the sticky nature of pollen allows it to adhere to the inside of nasal passages when people inhale, leading to an allergic reaction. An allergy is any severe immune response to foreign particles; an allergy to pollen is called *hay fever.*

Most areas in the United States have a local pollen-counting station that collects 35 cubic feet (1 m^3) of air over a 24-hour period. In a laboratory a technician then counts the magnified particles in a microscope. The unique pollen shapes allow technicians to estimate the amount of pollen belonging to different plant types, such as grass, ragweed, dogwood, and so on. The table on page 116 gives an example of a typical pollen count scale.

Hay fever, also called allergic rhinitis or pollinosis, ranges from a mild ailment to a serious health hazard. When people experience a mild allergic reaction to pollen, they may have temporary runny nose, sneezing, coughing, congestion, or a combination of these symptoms. Serious hay fever lasts year-round, saps a person's energy, and irritates the nasal passages. In moderate to serious allergies, sinusitis may develop in which the person's sinuses become inflamed. Asthma sufferers are especially at risk for complications from severe allergy.

Pollen allergies may get worse over time because of the way the immune system responds to foreign particles in the body. When the

POLLEN COUNTS		
POLLEN COUNT (NUMBER OF POLLEN GRAINS COUNTED IN 1 M^3 OF AIR OVER 24 HOURS)	CATEGORY	POTENTIAL HEALTH HAZARD
1–15	low	poses little health risk
16–90	medium	acceptable for most people, except those with chronic respiratory problems or allergy sensitivities
91–1,500	high	health alert in which almost everyone will experience some irritation or more serious symptoms
more than 1,500	very high	affects everyone, but hazardous to people with serious respiratory conditions; may lead to emergency medical care

immune system recognizes an *allergen* it has confronted before, it sets up a faster response than it did the previous time the allergen appeared. Unfortunately, in hay fever the body mistakes harmless pollen for a more dangerous particle and overreacts to the pollen's presence. In doing its best to protect the body from infection, the immune system actually can make a hay fever sufferer quite miserable. Hay fever has become a point of discussion within the larger issue of global warming due to changes in the growth of the world's plant life. This subject is explored in the sidebar "Case Study: Why Are Allergies Increasing Worldwide?" on page 117.

NOISE POLLUTION

Urbanization offers advantages that are not always available to people in rural settings. Urbanization has also brought a growing list of disadvantages to human and environmental health. Some of these factors are discussed in the case study "The Industrial Revolution" on page 120. The most serious disadvantages are the following: loss of habitat or habitat fragmentation; increased pollution; concentration of greenhouse gases; depletion of natural resources; increased volume of wastes; and

Case Study: Why Are Allergies Increasing Worldwide?

Doctors have noted an increase in allergies worldwide in the past several years, and they and ecologists have suggested the increased incidence is related to global warming. Plants thrive on carbon dioxide and the rising levels of carbon dioxide in the air have spurred the reproduction of plants and allowed trees to grow taller and fuller. The increase in plants and trees in urban areas results in more pollen. Physician Clifford Bassett, an allergy and asthma expert, mentioned to ABC News in 2007, "It's a super-mega pollen burst. Young people, elderly people, coming in and saying, 'Doc, I've never had this before. What's going on?'" Allergies have grown into a major part of general medical practices.

Between 1980 and the mid-2000s asthma cases have doubled. Pollen (combined with chemical air pollutants) has been implicated as the agent that aggravates the respiratory tract. Interestingly, food allergies have also been increasing. Theories abound on food allergies; one theory suggests that additives to food have increased the body's immune responses to hypersensitive levels. "Reduced fresh fruit and vegetable intake, more processed food, fewer *antioxidants,* and low intake of some minerals—these are all shown to be a risk," said Harold Nelson of Denver's National Jewish Medical and Research Center to *National Geographic* magazine in 2006. The relationship between food allergies and the environment has yet to be determined.

The debate most often returns to the environment and the mix of known and unknown things being put into the air along with more pollen. Though environmental medicine has not developed a theory that pleases every scholar, Bassett summed up the current thinking on the rise about allergies: "We're seeing more pollen production probably as a result of greenhouse gases, global warming." The spotlight seems to always return to environmental pollution.

generation of wasted heat energy. Water, air, and soil pollution are well-known problems, but cities also produce light and noise pollution. Light pollution interferes with the migration routes of birds and land animals and even interferes with aquatic life. For example, sea turtles come on shore only at night for laying their eggs on the beach. High light intensities that emanate from cities seem to prevent turtles from heading toward the shore because the light confuses their normal movements.

Noise pollution also causes subtle but troublesome effects in the environment. Noise that fills the air where people live and work is a health hazard for two main reasons: hearing loss and stress. Noisy areas also affect wildlife behavior. In humans, loud noise causes temporary hearing loss and sustained noise can lead to permanent hearing loss or impairment. Noise-induced hearing loss (NIHL) results from a combination of physical, genetic, and environmental factors. Loud noises damage the sensory cells lining the inner ear tube called the cochlea. Tiny hair cells that help sense sounds become distorted by sustained loud noise. They may then stiffen in this abnormal arrangement and thus become less effective at distinguishing sounds. NIHL is variable from person to person, but in general prolonged exposure to sounds louder than 85 decibels causes injury to the inner ear and possible hearing loss.

Sound loudness is measured in a unit called a bel (B); a decibel (dB) is one-tenth of a bel and the unit used for measuring the intensity of sound detected by the human ear. Bels and decibels are measured on a logarithmic scale to define the very wide range of sound intensities heard by humans. (A logarithm is a number expressed to a power; for example, 10^2 equals "10 to the power of 2" or 10×10.) An increase of 10 dB equals an increase in loudness of 10^1 or 10 times louder, and an increase of 20 dB or 10^2 is 100 times louder.

Sound frequency is measured in a unit called a hertz (Hz). Like loudness, frequency beyond a certain level can impair hearing. A person may experience hearing impairment exposed to frequencies nearing 4,000 Hz—normal speech ranges from 500 to 3,000 Hz.

Animals, including humans, evolved using hearing as a means of finding prey and sensing impending danger. Noise perception therefore relates to behavior that has been part of biological history. The constant din of noise in urban centers may interfere with these normal functions. Studies in England and Germany have showed that children attending schools near airports had higher levels of adrenaline and cortisol—the "fight or flight" hormones—in their blood, indicating that they lived in a constant state of physical stress. WHO environmental scientist Roberto Bertollini told the *Washington Post* in 2007, "This is the most disturbing thing about noise, because it means you are being exposed to this reaction all the time." Workers in jobs that expose them to high noise levels wear ear protection

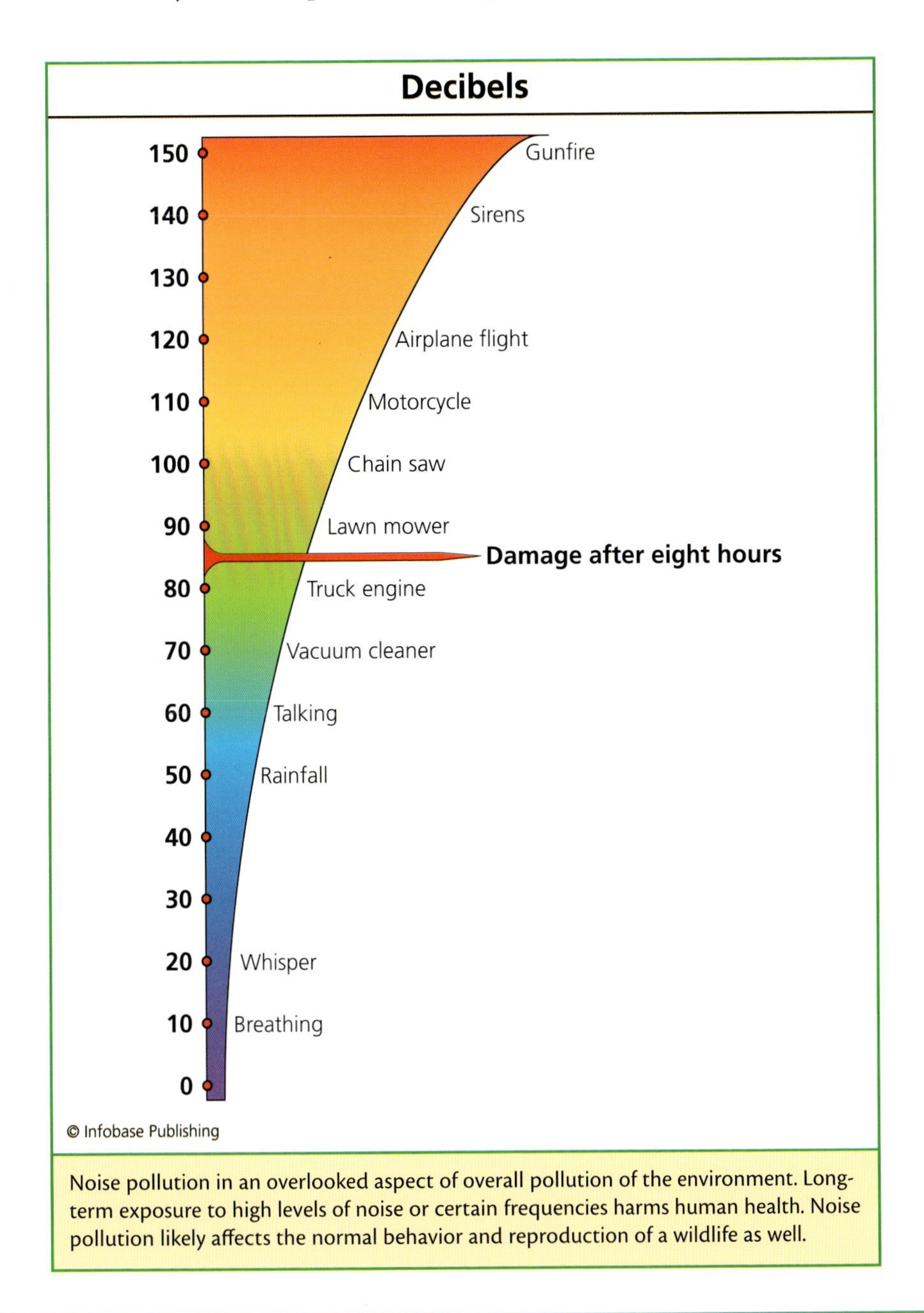

Noise pollution in an overlooked aspect of overall pollution of the environment. Long-term exposure to high levels of noise or certain frequencies harms human health. Noise pollution likely affects the normal behavior and reproduction of a wildlife as well.

CASE STUDY: THE INDUSTRIAL REVOLUTION

The Industrial Revolution brought many benefits to society and at the same time introduced the environment to its greatest threats. The dawn of human societies about 12,000 years ago subsisted by hunting and gathering foods. Settlements were small and each made use of a limited number of tools. Three major cultural changes took place to develop the industrialized world's present lifestyle: the agricultural revolution (10,000–12,000 years ago); the industrial and medical revolutions (beginning about 275 years ago); and the information-globalization age (starting about 50 years ago). The Industrial Revolution has received most of the blame for introducing environmental threats such as pollution, depletion of natural resources, and habitat destruction, even while it made people's lives easier.

The Industrial Revolution brought the use of factories that could mass-produce products for people living far away. This centralized production depended on trains, trailer trucks, and ships to transport goods. As transportation grew, so did air pollution, road construction, fuel production, and noise. Urban centers began growing with the rise of industrialization; rural farms took over undeveloped land and urban centers swallowed suburban areas. Commutes grew from a walk or a mile or two on a trolley to long drives. Long commutes led to clogged traffic near urban centers with more and more cars emitting carbon dioxide and additional greenhouse gases. As new technologies emerged, new greenhouse gases appeared, substances such as CFCs and VOCs.

The Industrial Revolution started about the mid-1700s in England and the mid-1800s in the United States. At the same time people endured crowed living and working conditions, long working hours, and inhaled smoke, soot, and other emissions in the air; infections and disease increased. By the 1960s environmentalists began to point to the rise of industrialization as the reason for the decline of the environment and perhaps human health. Eric McLamb wrote in 2008 for Ecology.com, "It was the fossil fuel coal that fueled the Industrial Revolution, forever chang-

and have their hearing checked annually, but people in cities usually do not take precautions to avoid constant noise. The easiest thing for people to do is to leave noisy areas as soon as possible to avoid prolonged exposure.

AIR POLLUTION REDUCTION TECHNOLOGY

Europe's air pollution condenses in an area known as the Black Triangle where Germany, Poland, and the Czech Republic meet. The Black Tri-ing the way people would live and utilize energy. While this propelled progress to extraordinary levels, it came at extraordinary costs to our environment and ultimately the health of all living things." Disease certainly did not begin with the Industrial Revolution, but it probably put a larger proportion of the population in health risk.

Before the Industrial Revolution, people's main health threat came from infections spread by vermin or a lack of sanitary conditions. Medical care improved during the Industrial Revolution and doctors began to gain an upper hand against infection. Meanwhile toxicities caused by smoke and chemicals increased. Today many people believe that the rates for cancer and other diseases increase directly with the amount of toxins in the environment. Not all of these relationships have been proved, but in 2003 physician Paul Kleihues said, "The [WHO's] *World Cancer Report* tells us that cancer rates are set to increase at an alarming rate globally. We can make a difference by taking action today." In fact, the WHO predicts that cancer rates will increase by 50 percent, equaling 15 million people, by the year 2020.

Rather than wait to prove all of the relationships between industrialization and disease, perhaps it would be more prudent stem the outpouring of toxins into the environment now. Kleihues and McLamb both hint that changes to the industrialized way of life may be called for to save environmental health. McLamb stated, ". . . . while coal and other fossil fuels were . . . taken for granted as being inexhaustible, it was American geophysicist M. King Hubbert who predicted in 1949 that the fossil-fuel era would be very short-lived and that other energy sources would need to be relied upon." Today, people recognize new energy sources such as wind, water, and solar power as a partial answer to managing environmental health threats. The new Industrial Revolution may be a sustainable revolution. By slowing use of natural resources and turning to renewable resources, society may lessen its health risks at the same time.

angle is one of Europe's most industrialized regions and for many years at least 10 coal-burning power plants and several manufacturing plants filled the skies with sulfur dioxide and other pollutants. People in the area suffered respiratory disease and cancer at rates above the rest of Europe, soot-covered forests died and deprived wildlife of habitat, and schools announced "smog days" to keep students home on the most polluted days. In 1987 electrical engineer Eduard Vacka described his hometown of Teplice in the Czech Republic: "It was one of those miserable fall days, when you wake up in the morning with a throbbing headache. Out the window, it looks like a dark sack has been thrown over the whole town. . . . God, what a stench! What the hell are they putting in the air? It's unbelievable: they're waging chemical warfare against their own people." In the past two decades the dire situation in the Black Triangle evolved from an environmental disaster to a cleaner and healthier place to live and work. How did this happen?

In 1990 the Czech Republic shook off the effects of the Soviet Union's domination. The Czechs began dismantling inefficient and dirty manufacturing plants and diversified their power sources with natural gas, hydropower (power generated from water), and nuclear power. Neighboring countries Germany, Hungary, and Poland received aid from the European Union to clean up their polluters as well. These countries tried a combination of approaches to reduce air pollution: employing emissions technologies such as desulfurization filters for cleaning emissions; closing down the small and most inefficient power plants and open-pit mines; and monitoring air quality in each country within the Black Triangle. The area also turned to clean coal technology, which enabled coal-burning power plants to release only emissions that have been cleaned of greenhouse gases and particles. Some groups tried a new device called an electrostatic precipitator, which removes tiny particles from emissions by capturing positively and negatively charged particles.

Countries in the Black Triangle also addressed air quality by asking industries to install pollution filters on manufacturing and power plants and improving mass transit. Czech historian Miroslav Vanek explained his country's progress to Radio Prague in 2005: "I think it was visible in the early '90s how quick the situation in northern Bohemia and all the Czech Republic got better and better . . . and you can see that now the situation in Teplice, for example, Ústí and other towns from North Bohemia is better, for example than Prague." The sulfur dioxide, nitrogen oxides, and particles in today's air over the Black Triangle have decreased more than 90 percent, 80 percent, and 95 percent, respectively since 1989. The Black Triangle illustrated that even very difficult air pollution crises can be reversed with cooperation between governments and residents.

CONCLUSION

Earth's lower atmosphere contains a multitude of gases and particles that have increased in density since the industrial revolution. These substances cause a variety of health hazards in people and animals. Many of the gases harmful to health are greenhouse gases, which are a major cause of global warming. The United States government and international agreements such as the Kyoto Protocol have put limits on the amount of greenhouse gases and fine particles that businesses may emit into the air. These actions have helped clean up the air in some regions but air pollution can travel to other places. Global air pollution truly requires the work of several countries at once to make a difference. The air pollution problem must be managed fast because scientists have shown that air pollutants have spread worldwide and airborne substances have been found in human and animal bodies.

Indoor air quality also presents a health risk, but offices and households have a better chance at improving it compared with outdoor air quality. Much indoor air pollution comes from common household items that emit irritating vapors. Homeowners can improve the air quality inside houses by storing chemical products safely, trying to purchase low-emitting products, and improving the home's ventilation.

The air seems to hold a bewildering array of substances that all cause health problems. People certainly would be justified in wondering how to deal with a diverse collection of substances such as carbon dioxide, particles, radiation, noise, and numerous other harmful pollutants. People can reduce the hazards of air pollution in several ways. First, strict government regulations on pollutants can improve air quality. Second, people could change behaviors that impact air quality, such as reducing their use of personal cars. Third, simply avoiding areas with obvious air pollution helps prevent respiratory irritation and more serious illnesses. The body can eliminate many inhaled substances and, provided that people steer clear of further contamination, they can return to good health. Some pollutants such as heavy metals, however, stay in the body for a long time and cause serious harm. Environmental medicine possesses a range of treatments that help rid the body of pollutants, but this area of medicine needs a good deal more study and the development of new treatments.

Poor air quality correlates with industrialized life. Though industry has brought benefits to society, society must now correct the damage it has done to the atmosphere. Science has developed an encouraging selection of technologies to clean up emissions and many local and national governments have instituted effective air pollution laws. Environmental medicine must do its part to study the health effects of air pollutants and add to the existing knowledge of hazardous gases and particles.

Food and Water Hazards

Air pollution has arisen with the growth of industry, but water pollution has occurred since the first human societies. Water has always been used by people for waste disposal and in many parts of the world, surface waters—oceans, bays, rivers, and lakes—remain a waste disposal site. Despite U.S. laws such as the 1972 Clean Water Act to protect the nation's waters, contaminated water remains a problem due to accidental discharges of wastes, intentional illegal dumping, and runoff.

Global water supplies range from very clean to very unhealthy. U.S. water utilities supply clean drinking water every day to millions of customers, but more than 1 billion people worldwide depend on intermittent and polluted water sources. Poor-quality water contains three main types of health hazards: toxic chemicals, infectious agents, and radioactive substances. The World Health Organization (WHO) has identified about 150 different chemicals in water known to be health hazards to people. Infectious agents consist of bacteria, viruses, cysts, protozoa, and parasites. These organisms tend to be high in waters contaminated with sewage but they can also be found in seemingly pristine waters. In wilderness areas where few people travel, wildlife wastes contaminate streams and lakes with organisms dangerous to humans. Radioactive contamination comes from two main sources: accidental release of radioactive materials and radon contamination of groundwater.

Civilization has always used flowing water as a convenient way to carry away wastes. Generations of people viewed the ocean as a place where wastes seemingly disappear or degrade. In truth, water dilutes wastes but it does not instantly degrade waste materials. Water pollution merely flows to a different location and spoils the clean waters in a new place. Contaminated water that stays in the soil can also pose a health risk by contaminating food sources. Foods, in fact, receive contamination from both polluted water and pesticides sprayed directly on them. Consumers must understand the potential health risks of water and foods and learn the best ways to assure good health.

This chapter covers the subject of water quality and food. The chapter opens by describing water-related health issues and the hazards in food that can cause illness. It also provides details on how contamination enters the body, specific food-borne and waterborne diseases, and the process of bioaccumulation of ingested pollutants. This chapter describes the current global state of safe water and food supply, water treatment technology, and the energy costs of producing food.

THE IMPORTANCE OF WATER QUALITY

Water is part of almost every chemical reaction in biology. If water is not part of a reaction, it likely provides the medium in which the reaction takes place. The human body contains 70 percent of its weight in water; cells other than bone cells provide a watery material called cytoplasm that fills most of the cell. Cells themselves contain 70 to 95 percent water. Tissues furthermore are bathed in blood, composed of about 55 percent plasma that is more than 90 percent water. The blood's water component carries nutrients into cells and wastes out of cells. Pollutants that travel through the body in blood, regardless of whether they came from drinking water or food, enter cells where the chemicals can disrupt several cell components: membranes, cytoplasm enzymes, deoxyribonucleic acid (DNA) inside the cell's nucleus, and protein activity. The nucleus is a specialized package inside cells that contains a person's genome.

The major functions of water in the body are the following: nutrient transport in blood; waste removal by blood; enzyme activity in the digestive tract; fluid for the lymphatic system; a solvent for chemical reactions in tissues; temperature regulation; and contribution to acid-base balance in the body.

People get most of their daily water by ingesting it; about two-thirds of intake comes from drinking water and other liquids and the remaining third from food. The body absorbs water through its intestinal lining in an energy-free process called diffusion. If the water or food has been contaminated with chemicals, this is the route these chemicals take to enter the bloodstream.

Despite the excellent water quality in the United States, some areas have room for improvement. Thousands of water supply systems may be in violation of the U.S. Environmental Protection Agency's (EPA) limits for at least one contaminant. These contaminants are likely to be substances that normal water treatment does not completely remove: volatile organic compounds (VOCs); the gasoline additive methyl tertiary-butyl ether (MTBE); some pesticides; nitrates; hormones; endocrine disrupters; and antibiotics. Water treatment does better at removing microbes, parasites, worms, large particles, and select elements such as arsenic, but even these things can sneak through treatment plants on occasion.

Water pollution is a much more serious problem in the developing world than it is in industrialized nations that have well-built water distributions systems. Poor sanitation and hygiene, inadequate controls against wastes in surface waters, and industrial discharges plague many

Regions in the world experiencing water stress have an increased vulnerability to disease. Surface freshwaters have receded in many parts of the world. Lake Mead in Nevada provides an example. The white band on the far shore in this picture shows that water levels have dropped, sometimes as much as 60 feet (18 m) in a three-year period. Snowmelt from the Sierra Nevada feeds Lake Mead, but global warming has decreased the snow levels, which affects the overall condition of water sources. *(Michael Molony)*

water sources worldwide. The World Health Organization's *World Health Report 2007* cites the following threats to water quality: epidemics of new diseases spread by water, toxic chemical spills, radioactive chemical accidents, and environmental disasters that contaminate water. These events can potentially ruin the quality of surface waters such as reservoirs, lakes, and rivers that serve as drinking water stores, and also groundwater sources, called aquifers, that supply well water. The WHO's 2008 report on world water quality, *Safer Water, Better Health,* states, "Almost one-tenth of the global disease burden could be prevented by improving water supply, sanitation, hygiene and management of water resources." This international health organization has identified the water-related illnesses listed in the following table as current or impending problems. All of the diseases listed arise from

Current Health Concerns from Poor Water Quality Worldwide

Infection	Type of Infectious Agent	Health Concern	Annual Effects
diarrhea	waterborne pathogens	dehydration, death, especially in children	1.4 million deaths
lymphatic filariasis (elephantiasis)	mosquito-borne parasite	appendages swell several times their normal size	25 million cases
malaria	mosquito-borne parasite	fever, vomiting, halted blood supply to organs	500 million cases, 1 million deaths
nematode infection	intestinal roundworms	infestation of vital organs	2 billion cases
schistosomiasis	snail-borne flatworm	severe urinary and intestinal damage	200 million cases
trachoma	fly-borne parasite	blindness	5 million visually impaired

Source: World Health Organization

CASE STUDY: THE ARAL SEA

The Aral Sea is a saltwater lake in Central Asia located between Uzbekistan and Kazakhstan of the former Soviet Union. People living near the lake depend on the desertlike land for growing cotton, rice, and other water-demanding crops. Since the 1960s water has been diverted from the Aral Sea for crop irrigation and now the lake's volume has decreased by 50–60 percent. A massive irrigation canal drew water from the Aral Sea for decades and left shallow pools behind, which evaporated quickly. The evaporation turned the remaining water more saline or salty. A farmer who lived his whole life in the area told the *New York Times* in 2008, "Thirty years ago, this was a cotton field. Now it's a salt flat." Fish and animals have succumbed to the altered ecosystem, but humans too have experienced health problems related to the Aral Sea.

The Aral Sea's health threats came from a series of events: (1) increased salinity in the shrunken water source affected drinking water quality; (2) farmers used large amounts of pesticides and herbicides to help crop yields in an unfavorable growing region; and (3) the dried lake bed served as a source of pollution-carrying dusts. These problems remain today and residents of the region contract respiratory illnesses, toxicities from dusts and salts, and toxicities from the water containing increased concentrations of chemicals.

The long-term health of the Aral Sea communities can be improved by stopping the pollution rather treating illnesses year after year. Since 1999 the United Nations and the World Bank have spent several million dollars on projects to make irrigation more efficient, capture drainage

infectious agents, such as parasites. The sidebar above "Case Study: The Aral Sea" describes a situation in which a threatened water supply indirectly affects community health.

HEALTH HAZARDS IN FOOD

Contaminants enter food chains in the following five ways: (1) pesticide use; (2) infectious pathogens in irrigation water; (3) infectious pathogens from wildlife wastes in cultivated fields; (4) contamination with chemicals or pathogens during handling or processing; and (5) contamination with chemical toxins or infectious agents during preparation.

Cooking eliminates infectious organisms in food even if the food had been improperly handled or processed. Improper cooking—too little

water, and restore wetlands that contribute to purifying the water and providing animal habitat. A few farmers have replaced high-water crops such as cotton and rice with crops that do well with less water, such as wheat.

The draining of the Aral Sea has left a generation of women with poor health caused by anemia. High levels of the metals zinc and manganese in the local water interferes with iron absorption, which in turn interferes with the normal functioning of oxygen-carrying hemoglobin in the blood. The Aral Sea region also has high incidences of kidney and thyroid disease and throat (esophageal) and liver cancers. Crops for food present as big a health threat as the water because farmers often flood their fields to hold in moisture. By doing this, they put a pesticide coating on food crops.

In 2004 BBC News reported that people in the Aral Sea region had a high incidence of damage to their DNA and also high cancer rates. By 2007 communities (funded by the World Bank) had built levees and dams to retain the Aral Sea's water and the lake has made a modest recovery, but the long-lasting health effects may not disappear as easily. Said the World Bank's project leader Joop Stoutjesdijk to Nature News in 2007, "The impact has been very swift and very important for the local people, both from a health point of view and from a livelihood point of view." But he also warned, "You have to be careful of being overly optimistic. If you ask, 'Are they saving the Aral Sea?' the answer is clearly no." The Aral Sea's ongoing dilemma illustrates how water scarcity leads to a domino effect of health problems.

cooking time or too low a temperature—allows pathogens to remain alive to cause food-borne illness. Some pathogens produce toxins so that even if cooking kills the pathogen, the toxin stays. Sanitary handling of food and proper cooking are the best ways to make foods safe from pathogens.

Cooking does not affect chemicals and in many places in the world, the following materials can be found in food: pesticides; mercury; persistent organic compounds (POPs) such as dioxins, polychlorinated biphenyls (PCBs), and pentachlorophenol (PCP); flame retardants such as polybrominated diphenyl ethers (PBDEs); and residual antibiotics or growth hormones in poultry and beef. Some foods also carry natural toxins. Many of these natural toxins come from molds—toxins made by molds are called mycotoxins—that grow on specific foods, as the following examples show:

- fumonisins from *Fusarium* molds that contaminate corn
- patulin from *Penicillium* and *Aspergillus* molds that contaminate apples
- aflatoxin from various molds that contaminate grains
- ergot from *Claviceps* mold that contaminates wheat, rye, barley, and oats
- deoxynivalenol from *Fusarium* that contaminates various grains

The type of food a person eats determines the type of toxin that he might ingest. Some farms feed animals antibiotics and growth hormone that are both thought to improve meat and milk production. The United States approves the following three hormones to improve the efficiency of meat production: (1) the female hormones estradiol and progesterone, (2) the male hormone testosterone, and (3) the synthetic growth-promoting hormones zeranol, trenbolone, and melengestrol acetate. The production process also contaminates meat, milk, and eggs with dangerous bacteria that must be killed by cooking or pasteurization. Crops such as fruits and vegetables, by contrast, have a higher risk of contamination by pesticides, chemicals in irrigation waters, or pathogens in the wastes from wildlife that spend time in cultivated fields.

Seafood also receives contamination from pathogens and chemicals. Bottom-dwelling mollusks have been contaminated with sewage runoff and pesticides. Heavy metals have been especially troublesome in other seafood species. The U.S. Food and Drug Administration (FDA) lists tilefish, swordfish, shark, and king mackerel as the ocean fish with the highest mercury contamination.

CONTAMINATION ROUTES

Food and water toxins get into the body mainly by ingestion, through the skin by way of cuts or by absorption, and by inhalation. Exposure to polluted recreational waters at beaches or rivers may result in a person receiving the toxins by unintentional ingestion. On rare occasions, bath or shower water can transmit environmental toxins. In these cases, a person may take in the toxin by inhaling tiny moisture particles called aerosols or through an open cut or wound.

Workers in the fishing industry also must be aware of any skin lesions or neurological disorders that appear after coming in contact with water. These symptoms are the hallmark of toxins made by algae. In the 1990s a type of alga called a dinoflagellate caused serious health problems among people working in the Chesapeake Bay and along North Carolina's rivers. This organism named *Pfiesteria piscicida* represented only one of almost 100 algal species that harm human, animal, or aquatic life, and one of more than 1,000 algae called *dinoflagellates,* but its unique behavior made it a notorious example of pollution. When fish enter *Pfiesteria*-infested waters, the microbe detects the fish's presence and millions of these organisms swarm toward the fish. This occurrence kills the fish quickly and in large numbers, an event called a fish kill. Fishers have suffered the following ailments when coming in contact with infested waters: shortness of breath, nausea, memory loss, and skin sores similar to those found on infected fish. In 1997 when the *Pfiesteria* contamination reached its height, a local doctor had remarked that he had treated a man who "suffered severe headache and thirty lesions after waterskiing in the lower Pocomoke [River, Maryland] for a half-hour." Researchers eventually solved the mystery of *Pfiesteria* by studying the epidemiology of the illnesses. Public health departments in Virginia, Maryland, and North Carolina continue to monitor the ongoing threat of *Pfiesteria* blooms in the Chesapeake Bay. This event demonstrated how environmental health hazards can catch a community unprepared to fight them.

The *Pfiesteria* event also showed the role of media in stirring alarm in a population that at the time had no idea what was happening to cause thousands of fish to die suddenly. Writer Michael Fincham recounted to the *Chesapeake Quarterly* in 2007 that the *Pfiesteria* hazard was real, but perhaps overstated: "In hindsight, it's clear those [news] stories, through their sheer volume, exaggerated the risk posed to fish, people, and the environment. They led to mass panic and major economic loss." People continue to worry over real or perceived dangers of environmental toxins. Good information and sound advice from medical doctors can help the public navigate incidents such as this one.

Toxins in natural waters can cause massive deaths of fish species, called fish kills, which can involve many thousands of fish at a time. Fish killed by *Pfiesteria* have red lesions on their skin. People also suffer toxic effects from *Pfiesteria*, mainly neurological disorders. *(U.S. Fish and Wildlife Service)*

FOOD-BORNE AND WATERBORNE ILLNESSES

A food-borne illness is caused by any agent, biological or chemical, that enters the body with food. Waterborne illnesses are biological or chemical substances that enter the body with drinking water, recreational waters, or exposure to other contaminated surface waters. Scientists can have a hard time pinpointing the source of outbreaks from either food or water for the three following reasons: (1) millions of cases may go unreported, especially in developing countries; (2) an additional percentage of illnesses are never identified; and (3) outbreaks can be sporadic and go undetected. Food or water illnesses cause these problems because contamination in them acts in an unpredictable manner until scientists can make sense of the outbreak. The main food-borne threats as identified by the WHO are listed in the following table. Some of these illnesses may decline over time and others take their place as the nature of pollution changes and as environmental medicine advances.

Major Food-borne and Waterborne Illnesses			
Illness	**Cause**	**Main Sources**	**Symptoms**
Infectious Agents			
salmonellosis	*Salmonella* bacteria	eggs, poultry, dairy products	fever, headache, nausea, vomiting, abdominal pain, diarrhea
campylobacteriosis	*Campylobacter* bacteria	milk, poultry, drinking water	abdominal pain, fever, nausea, diarrhea
enterohemorrhagic fever	*E. coli* bacteria, viruses	meat, fruits, vegetables	severe abdominal pain, diarrhea, fever
cholera	*Vibrio* bacteria	water, seafood	abdominal pain, vomiting, diarrhea, dehydration
cryptosporidiosis	*Cryptosporidium* protozoa	drinking water	severe abdominal cramping
Chemicals			
various toxicities	mycotoxins	dried foods	neurological damage
persistent organic pollutants (POPs)	industry, incinerators	foods, water	neurological and other organ disease, cancer
mercury poisoning	mercury contamination	seafood	neurological damage, birth defects
arsenic poisoning	minerals, rocks	drinking water	vomiting, throat and abdominal pain, diarrhea, lung cancer, bladder cancer

Clean water without any chemical or biological toxic substances is difficult to find in many parts of the world. Industrialized countries enjoy available and safe tap water. Other countries do not have the same luxury. These women in Uganda carry water from a distant water source to their homes. *(Kadami Hospital)*

Some chemicals that are natural in the environment end up in drinking water sources like lakes, reservoirs, or rivers. Geology, soil chemistry, and runoff patterns determine the amounts of chemicals in surface waters. Some natural contaminants that can cause occasional health problems are arsenic, barium, boron, fluorine, uranium, manganese, and molybdenum.

In 2007 the environmental advocacy organization U.S. Public Interest Research Groups published a press release on U.S. water quality: "More than 57 percent of industrial and municipal facilities across America discharged more pollution into our waterways than their Clean Water Act permits. . . ." The group's spokesperson, Christy Leavitt, mentioned in the press release, "As the Clean Water Act turns thirty-five, polluters continue to foul our rivers, lakes and streams. With so many facilities dumping so much pollution, no one should be surprised that nearly half of America's waterways are unsafe for swimming and fishing." The United States can certainly improve its water and food quality as Leavitt suggests, but this country remains an example of providing a healthy and clean diet compared with many other parts of the world.

BIOACCUMULATION

Bioaccumulation is a gradual increase in the concentration of a toxin in the tissues of a living thing. Some toxic substances persist for long periods in the body for one of two reasons. First, the substance dissolves in the body's fat stores and does not mobilize from there. Compounds with large structures and one or more *carbon rings* tend to infiltrate fatty tissues and accumulate there. Second, tissue enzymes cannot degrade the toxins fast enough to keep up with additional ingested toxins.

Bioaccumulation does not continue in a limitless fashion, and indeed the body degrades and excretes toxins as fast as it can. At first, tiny amounts of toxin move easily into cells and the toxin concentration increases. Eventually, the concentration of the toxin reaches a point in which its entry rate into the body equals its excretion rate from the body. With chronic exposure to a toxin, the body cannot degrade the chemical fast enough to prevent bioaccumulation and then toxicity. A person can rid himself of many types of toxins by moving to a clean environment with no further exposure to hazardous substances. Over days or months the liver detoxifies the chemical and the body excretes it.

MEASURING FOOD AND WATER QUALITY

Food or water suspected of contamination with infectious agents may be tested using microbiology. Scientists count relatively large things such as worms, parasites, protozoa, and cysts in a microscope. Microscopic particles such as bacteria require more involved methods in which the scientist dilutes a measured portion of food or water and transfers it to a medium (a source of nutrients for microbes). Following incubation, visible colonies of bacteria may be counted in a microscope. In water and food quality testing, the presence of large amounts of bacteria is equated to large numbers of other microbes such as viruses.

For chemical analysis, laboratories contain sensitive instruments that detect pesticides, organic compounds, allergens, mycotoxins, particles, and heavy metals. In the United States, water testing laboratories measure specific constituents to determine if the water is safe for use as drinking water. Fresh foods and meats usually receive less testing. The table on page 136 summarizes the constituents measured by water laboratories. After

Example Drinking Water Quality Report

Constituent	Units	Main Source
Health Related		
fluoride	mg/l	erosion of natural deposits
nitrate	mg/l	runoff containing fertilizers or leaching from septic systems and sewers
radioactivity, alpha particles	pCi/l	erosion of natural deposits
chlorine	mg/l	an added disinfectant
copper	µg/l	corrosion of plumbing systems
lead	µg/l	corrosion of plumbing systems
trihalomethanes	µg/l	disinfection by-product
haloacetic acids	µg/l	disinfection by-product
radon	pCi/l	granite soils
coliform bacteria	% positive samples	naturally present in environment
Aesthetics Related—Taste, Odor, Clarity		
color	spectrophotometer units	naturally occurring organic matter
odor	threshold odor number (TON)	naturally occurring organic matter

CONSTITUENT	UNITS	MAIN SOURCE
turbidity (cloudiness)	nephelometric turbidity units (NTU)	soil runoff
dissolved solids	mg/l	runoff, leaching of natural deposits
hardness (as calcium and magnesium)	mg/l	leaching of natural deposits
conductance	µmhos/cm	substances that form ions and so conduct current in water
chloride taste and odor	mg/l	runoff, leaching of natural deposits
sulfate odor	mg/l	leaching of natural deposits
sodium taste	mg/l	naturally occurring and from treatment method
manganese taste	µg/l	leaching of natural deposits
mg/l = milligrams per liter; µg/l = micrograms per liter; pCi/l = picocuries per liter; µmhos/cm = micromhos per centimeter		

the laboratory completes these analyses, the municipal water utility makes the results available to the public in the form of a water quality report, shown above.

Today most water utilities test for other unhealthy constituents that may be present in water. Laboratories check for the presence of the MTBE, which affects taste and odor and is a possible health threat. Most water laboratories also monitor arsenic levels and check for a protozoal cyst called *Cryptosporidium* that has caused outbreaks worldwide.

Water and food analysis laboratories rely on an assortment of equipment to determine the amounts of constituents in samples. The most common instrumentation for these analyses is listed here:

- atomic absorption spectrophotometer—detects elements by analyzing flame color
- gas chromatograph—detects volatile organic compounds
- gas chromatograph/mass spectrometer—measures organic compounds by flow rate in a gas and analysis of constituent chemical units
- infrared spectrometer—measures specific chemicals
- scintillation counter—measures alpha particle radioactive decay
- nephelometer—measures light impedance due to cloudiness
- biological oxygen demand (BOD) unit—measures dissolved organic matter microbiologically
- conductivity meter and probe—measures electrical conductivity

Water and food analysis laboratories supplement their sensitive equipment with more hands-on chemical and microbiological testing. Chemical testing done by hand using glassware and chemical solutions is called wet chemistry. Food and water testing therefore combines cutting-edge analyses with less technical laboratory methods.

WATER TREATMENT

Water treatment plants use a stepwise process to remove suspended matter, organic materials, and microbes from water so that the water is safe to drink. But while these plants remove certain substances such as small particles and pathogenic bacteria, the treatment might leave other constituents in the water. VOCs, for example, often escape the treatment plant.

Typical water treatment relies on four main processes to clean the water, as follows: (1) filtration through screens and small-pore filters to remove particles; (2) settling, which allows fine solid matter to drop to the bottom of a settling tank by gravity; (3) biological digestion of organic compounds with bacteria; and (4) disinfection with chlorine, chlorine-containing compounds, ozone, or ultraviolet light. Water flows through a water treatment plant on a route that allows these four steps to be their most efficient. Water treatment is not always foolproof, however, and two occurrences can threaten a community's health: (1) large rainstorms that overrun a treatment plant and (2) *disinfection by-products* (DBPs).

In 1993 a series of heavy rains pounded Milwaukee, Wisconsin, flooding surrounding farmland and creating excess runoff that entered streams and rivers. Storm drains soon overflowed and water treatment plants could not clean the inflow fast enough. *Cryptosporidium* cysts that filters usually catch passed through the plant and entered drinking water. More than 400,000 people became sick with cryptosporidiosis, an abdominal infection causing severe *gastroenteritis* and pain; about 100 individuals with weakened immune systems died. Milwaukee resident Gary Wells told CNN after the outbreak ended, "I used to drink this [tap water] because I thought I could trust that it was OK. But obviously I was wrong." Several other *Cryptosporidium* outbreaks have occurred since Milwaukee's experience. For this reason, during heavy rainstorms and disasters such as Hurricane Katrina, residents should listen to broadcast emergency alerts from a public health department to learn if their water is safe to drink.

DBPs present a chemical hazard, not a microbial threat like *Cryptosporidium.* DBPs form during the reaction of a disinfectant with naturally occurring organic matter in the water. The type and amount of DBPs formed depends on three aspects of water treatment: (1) the type of disinfectant; (2) the disinfectant dose; and (3) any disinfectant left over after killing the water's microbes. DBPs are known to arise from chlorine-containing disinfectants and ozone.

Two types of DBPs affect human health: trihalomethanes and haloacetic acids. Trihalomethanes include chloroform, bromodichloromethane, dibromochloromethane, and bromoform. Trihalomethanes have the following generic formula, where X is a usually chlorine or bromine:

$$X_3CH_3$$

Haloacetic acids of concern in water treatment are: monochloroacetic acid, dichloroacetic acid, trichloroacetic acid, monobromoacetic acid, and dibromoacetic acid. The formula for dichloroacetic acid, for example, is the following:

$$CHCl_2COOH$$

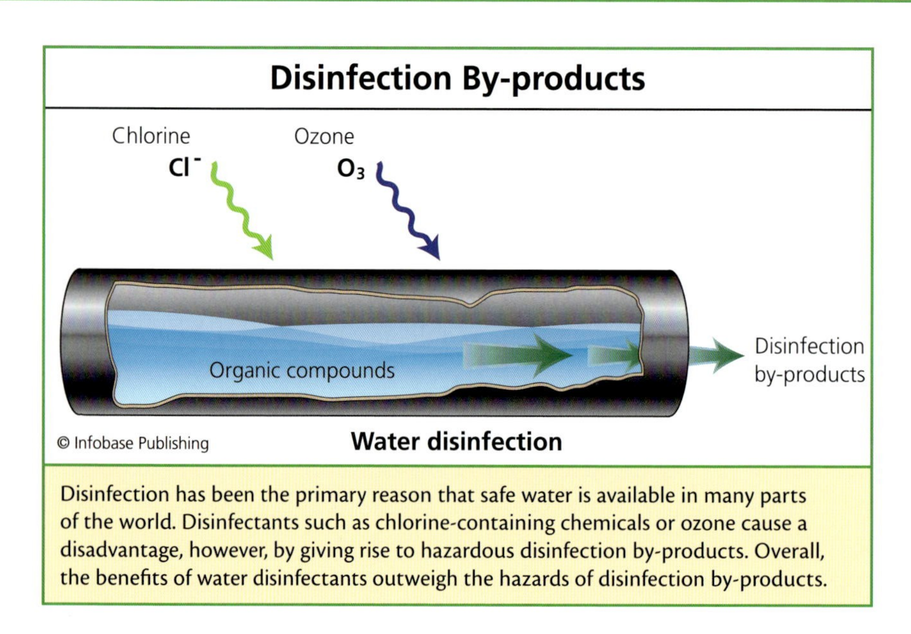

Disinfection has been the primary reason that safe water is available in many parts of the world. Disinfectants such as chlorine-containing chemicals or ozone cause a disadvantage, however, by giving rise to hazardous disinfection by-products. Overall, the benefits of water disinfectants outweigh the hazards of disinfection by-products.

DBPs have been suspected of raising the risk of bladder cancer and anal cancer based on laboratory data. Many questions exist on the safety of DBPs, but as a precaution, the EPA enforces the Disinfectants and Disinfectant Byproducts Rule of 1998. This rule sets upper limits of certain compounds in water known to form DBPs: chlorine, chloramines, chlorine dioxide, bromate, chlorite, the trihalomethanes, and haloacetic acids.

The water treatment industry has experimented with alternative disinfectants to avoid forming DBPs altogether. In 2004, for instance, Corpus Christi, Texas, tried chloramine disinfection to replace DBP-producing chlorine. Instead of solving the DBP problem, however, the water contained new DBPs that scientists feared could be more dangerous than the original chemicals. University of Illinois toxicologist Michael Plewa noted, "This research says that when you go to alternatives, you may be opening a Pandora's box of new DBPs, and these unregulated DBPs may be much more toxic . . . than the regulated ones we are trying to avoid." The puzzle of DBPs continues, so the public and the EPA must lean toward extra caution until all the health effects of these compounds become known.

Wildlife must suffer the consequences of chemicals and pathogens in waters they drink. Fish also confront all of these hazards plus physical changes that sometimes occur in water, highlighted in the following "Thermal Water Pollution" sidebar.

Thermal Water Pollution

Thermal pollution occurs when a large volume of heated water enters lakes, rivers, or coastal areas, causing the ambient water temperature to change several degrees. (Ambient water is the natural water in an organism's immediate surroundings.) Most of this heated water comes from industrial plants and electric power plants—almost half of all the water drawn from surface sources in the United States is used for cooling equipment in electric power plants. Heat-generating reactions or equipment in these facilities need cooling water to prevent overheating, and natural waters provide an inexpensive choice.

When cooling water finishes circulating around equipment its temperature has risen several degrees. The sudden discharge of hot water into natural waters causes significant damage to aquatic life. Heated water decreases the amount of dissolved oxygen in natural waters because much of its oxygen dissipates in the heating process. The decreased oxygen levels and increased temperature disrupt food webs and kill fish by an event called *thermal shock*. This phenomenon also occurs when wind patterns and currents cause ocean upwellings of cool water. Regardless of the source of unusual temperature waters, aquatic organisms suffer because they have evolved with enzymes that work only in a certain temperature range. Aquatic animals often die from thermal shock, and fish that manage to receive a mild shock and stay alive may develop reproduction disorders.

Power plants have the responsibility to eliminate thermal pollution by using one of three methods. First, the power plant can use cooling towers or cooling ponds where heated water remains until its temperature returns to normal. Cooling towers circulate the heated water until it cools, dissipating the heat into the air. Cooling ponds simply allow some of the heated water to evaporate as the rest of it cools. Second, the plant can build a dilution system that feeds cold water into the same pipes used by the heated water. As the water flows in the pipes to the outlet, the temperature decreases close to natural water temperatures. Third, the plant can release heated water slowly rather than in a large, sudden discharge, which allows fish to sense the temperature change and escape to more comfortable waters; this method is called the fish-chase method.

(continues)

(continued)

Thermal pollution opens the door to invasive species that thrive in warm water. Many invasive species from other parts of the world use aggressive tactics to drive native species out of their natural habitat. Without adoption of one of the above methods, thermal pollution can decrease the biodiversity of aquatic ecosystems near industrial or power plants.

GLOBAL FOOD DISTRIBUTION

Methods for distributing food around the world today contribute in part to food contamination. Tea, salt, and spices have traveled the oceans since antiquity, but people usually grew their staples at home. Until the 1940s to 1950s in the United States, fruits, vegetables, milk, and meat came from local farms. In the 1960s people began to accept a widening variety of canned foods and packaged meals and, before long, food production systems had also changed. Meat and poultry products now come from massive slaughter and processing plants, mostly in the western plains states and the southeastern states, respectively. Meat processed this way pools together several thousand animals and the meat products then travel by truck to grocery markets half a continent away. Crops follow a similar long-distance route from harvesting, processing, and packing to markets. Global shipping has become so efficient that little food spoils before it arrives at its destination.

About 25,000 food shipments arrive in the United States daily from all over the globe. Fruits and vegetables from Mexico, Central America, and South America give shoppers more choice and year-round fresh produce. The FDA holds the responsibility for monitoring the safety of the country's food supply, yet the deluge of imported goods has swamped the FDA's capabilities. In 2007 William Hubbard, a former FDA associate commissioner, told *USA Today,* "The FDA has so few resources, all it can do is target high-risk things, give a pass to everything else and hope it is OK. The public probably has the perception . . . that they're more protected than they really are." These are not comforting words considering that many countries use larger amounts of pesticides than the United

States and many of these pesticides have been banned in the United States. FDA lawyer Benjamin England said at the time that Hubbard's assessment had probably been accurate and that the only reason the public had not grown concerned was its good fortune of seeing relatively few food poisoning outbreaks.

Globalization of economies and international trade has helped food move around the world. An individual farm's economics also goes toward the decision of whether a crop will be mainly domestic-grown or foreign-grown. *USA Today* quoted Frederic Hauge of the Norwegian environmental group Bellona: "Food is traveling because transport has become so cheap in a world of globalization." But the beginning-to-end task of producing food lacks efficiency, as explained in the "Energy Costs of Food Production" sidebar.

Energy Costs of Food Production

Food production, particularly animal products, comes at a cost. Despite the efficiencies offered by mass production, animal products remain a very inefficient way to convert energy into a form that can be used by humans. Animal protein requires about eight times more fossil fuel energy to produce than the same amount of plant protein. The following list provides the energy input to protein output for different animal products: beef, 54:1; lamb, 50:1; eggs 26:1; pork, 17:1; and milk, 14:1.

Grain-fed cattle for beef production also use up 12,011 gallons of water for every pound (100,000 l/kg) of food produced; broiler chickens use 420 gallons/pound (3,500 l/kg). Vegetable production, by comparison, is more water-efficient: rice uses 230 gallons/pound (1,912 l/kg); wheat uses 108 gallons/pound (900 l/kg); and potatoes use 60 gallons/pound (500 l/kg).

Though oceangoing ships consume large amounts of fossil fuel, they carry tons of food in each shipment, which helps make each trip efficient. Many ships have been refitted recently to use cleaner-burning engines, but thousands of ships crossing the ocean every day still adds up to large carbon dioxide emissions. Unlike cars, woodstoves, or even lawn mowers, few regulations control air pollution from cargo ships. Because cargo ships have registrations from a variety of countries, pollution regulations on them have remained difficult to enforce.

MANAGING THE WORLD'S SAFE WATER SUPPLY

Safe water supply has been a difficult goal to attain in many regions of the world. Millions of people must find water anyplace they can, even water that contains toxic chemicals or pathogens. The WHO in its 2008 report, *Safer Water, Better Health,* stated that three different focus areas should be addressed to clean up the world's water supply: (1) water suppliers that distribute water in a way that prevents infectious disease and exposure to toxic chemicals; (2) an emphasis on improved hygiene and sanitation; and (3) improved management of activities to prevent pollution at its source.

In regions of the world where distribution systems are absent, people walk miles to collect water from the nearest well, and then carry the load back to their village. Water projects may be hampered by lack of money, political conflicts, or distance and terrain, so these ancient methods of retrieving water continue. The WHO therefore suggests that in-home water purification be installed prevent waterborne health threats, com-

The seawater desalination plant in Perth, Australia, is the largest in the Southern Hemisphere. Desalination is expected to become increasingly critical as a source of safe, clean water. *(ABB Engineering)*

bined with sanitation and good personal hygiene. Clean water, sanitized buildings, and regular washing reduce the chances of taking a toxin into the body. Governments, too, must consolidate the various water needs of agriculture, industry, and domestic use in a way that leaves water for natural ecosystems. Money would help solve some of these problems. For example, local communities that have never known anything but outhouses for disposing of wastes might be able to build a sewer system with a small amount of financial aid.

CONCLUSION

Water quality has greatly improved in industrialized nations in the past century, but water pollution continues to threaten people's health in many developing parts of the world. The importance of water to life on Earth cannot be overstated; every living thing depends on it and water makes up most of all biological cells. But polluted water enters the body as easily as clean water and when it does it brings toxic chemicals with it. A significant portion of the illnesses in developing countries likely comes from tainted water and food. The need for safe water and food has grown into a priority for world health concerns.

Protecting the global food supply and local drinking water sources requires a blend of science, sanitation, and hygiene, in addition to dedicated government involvement. The following future technologies that will be called upon to improve water and food quality: sensitive analytical detection of contamination; effective and safe compounds for water disinfection; and education in hygiene to reduce ingestion of health hazards. All of these things must receive support from governments to protect their citizens' health.

6 Populations at Risk

Environmental hazards assault healthy people every day, but a human body contains systems to prevent long-lasting harm. Skin provides a physical barrier against pathogens and toxins, the immune system destroys any infectious agents that enter the bloodstream, and the liver detoxifies chemicals to be excreted. It all works well. Even people suffering from a cold can battle other threats and return to good health.

A portion of the world's human population has lost much of this capacity to ward off environmental toxins. This subpopulation is called an at-risk or high-risk group. The following groups have greater than normal risks for disease, including toxicities from environmental factors: infants and very young children; the elderly; diabetes patients; pregnant women; persons with weakened immunity (organ transplant patients, AIDS patients, people with autoimmune disease); cancer patients; and people with an existing serious disease or injury. In these cases an environmental toxin has an increased chance of causing harm to the body.

Scientists additionally consider risk-informed or risk-based decisions. Risk-informed decisions operate in situations in which science holds partial information about a risk. For example, certain species of fish have been analyzed and found to contain mercury in their tissues. Health officials can warn the public against eating too much of this seafood based on this knowledge. Risk-based decisions must be made with no scientific information at all regarding a given risk. For example, a fisherman may walk to his favorite fishing spot on a wharf, but when he views the water he sees garbage or perhaps a dead fish. The fisherman makes a risk-based decision on whether to fish at that spot.

Risk is therefore part of every life, though people seldom calculate their risks consciously. Risk equals the possibility of something bad occurring; risk assessment is the process of determining the chances of that bad occurrence. In business, risk specialists assess risk with mathematical equations and statistical tools such as probability. People outside business perform the same process by using their intuition. For example, a student wishes to cross the street but the light is red. She may assess the traffic, approximate when the light will turn green, and calculate her risks of stepping off the curb before the light changes. This risk assessment tells the student the likelihood of crossing the street without being hit by a car. Wildlife also assesses risk as a survival tool. If a predator comes upon a prey animal, it must assess the size and strength of the prey to calculate its chances of capturing and killing its next meal. Sometimes a predator will like its chances and attack; other times it will turn away in search of easier prey.

Health agencies use risk assessment to determine the dangers of different environmental toxins. The risk of disease relates to the exposure period and the toxin's dose. Because science has not yet determined the details of how each environmental toxin affects every type of living tissue, agencies such as the U.S. Environmental Protection Agency (EPA) calculate risks of harm from toxins based on the knowledge at hand. *Regulated chemicals* are chemicals put into the environment by humans and known to cause harm to people or animals. The EPA regulates them by setting upper limits of these chemicals in the environment and uses either risk-informed or risk-based decisions to do this. In other words, U.S. residents cannot expect to live in a place that is 100 percent free of all contamination, but the government tries to reduce the chances of illness by assessing risk.

Risk assessment consists of four basic steps: (1) hazard identification, which involves defining the hazard and its known effects on human health; (2) exposure assessment, which is the determination of a toxin's concentration in the environment and the dose that enters a person's body; (3) dose-response assessment, which describes the likely health effects due to exposure to a toxin; and (4) risk characterization, which estimates the potential severity of illness due to the dose of the toxin.

This chapter explores the risks from environments that may contain environmental toxins. The chapter opens with a discussion on population growth and its relationship to environmental risk, then describes how the composition of a population influences disease rates. Sections on age, hazardous site jobs, the risks of living near toxic sites, and the medicine that treats at-risk people are also covered.

POPULATION GROWTH

Population growth has changed the health risks that confront people today. In the 1800s settlers moved west to find clean open space for ranching, growing crops, and raising a family. In the 21st century families have far fewer options for escaping environmental toxins and urban encroachment. Environmental risk may no longer be a choice between clean rural life and polluted urban life, but it seems to have become a choice between the risks of different types and levels of toxic exposures.

The population explosion since the Industrial Revolution has also created a more troubling and perhaps controversial dilemma. That is, the human population no longer abides by nature's rule that the fittest individuals earn the greatest likelihood of surviving. Wildlife, aquatic life, and plants develop according to the "survival of the fittest," meaning the individuals best adapted to their environment have a greater chance to live to reproductive age so that their genes pass to their progeny. Human society, by contrast, developed medical advances and other sciences that enable infants with serious diseases or birth defects to survive, and drugs that allow the infirm to live to an older age than would be expected without medical help. Individuals with a lower than average intelligence also have as much right to a healthy, happy life as individuals with higher than average intelligence. As a consequence the human population has an increased overall risk of poor health. In 2007 London's *Telegraph* contained an article on the relationship between the Industrial Revolution and how societies developed thereafter. One passage explained, "In pre-industrial societies women typically had five children. If living standards were good, most of those children survived to adulthood and rapid population growth followed. But with limited resources, only two of those five children survived to adulthood and the population remained stable." This theme has been explored by authors such as economist Gregory Clark in his 2007 book *A Farewell to Alms* and geography and physiology professor Jared Diamond's 1997 Pulitzer Prize–winning book *Guns, Germs, and Steel.*

Wildlife demonstrates the relationship between population growth and survivability. In wildlife, the well-being of a population can be determined by measuring the following seven traits: juvenile survival, adult

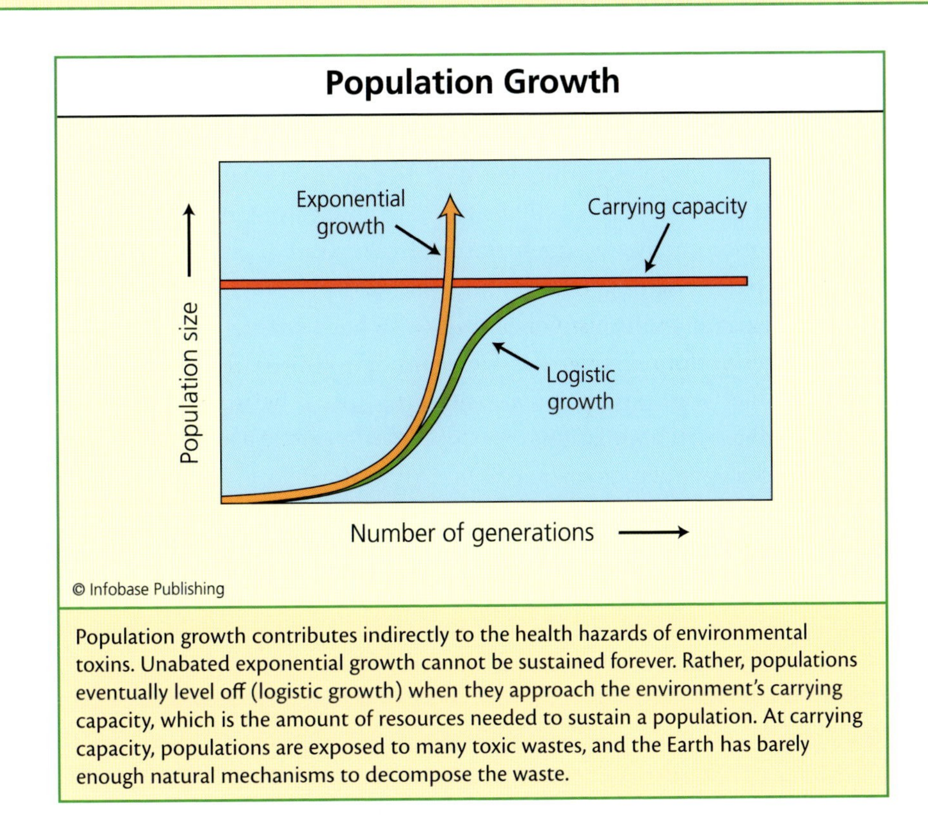

Population growth contributes indirectly to the health hazards of environmental toxins. Unabated exponential growth cannot be sustained forever. Rather, populations eventually level off (logistic growth) when they approach the environment's carrying capacity, which is the amount of resources needed to sustain a population. At carrying capacity, populations are exposed to many toxic wastes, and the Earth has barely enough natural mechanisms to decompose the waste.

survival, time to first reproduction, time between broods, total number of broods per individual, total number of offspring per individual, and population growth rate. Population growth rate equals the annual percent change in the size of a population, as shown in the following example, where V is population size:

- Growth rate = $[(V_{Present} - V_{Past}) \div V_{Past}] \times 100$
- 1998 population size = 100,000
- 2008 population size = 117,000
- Population growth rate = 17 percent

Population growth limits itself because of density-independent factors and density-dependent factors. Weather and climate are independent of population density and cannot be controlled by the population (although a person can rightly argue that global warming is controllable). Weather and climate therefore represent density-independent factors. The density of a population of people or wildlife, on the other hand, affects such density-dependent factors as food, water, and shelter as well as disease. Environmental toxins influence the incidence of disease, so a relationship can be said to exist between population density and environmental disease.

The relationship between population growth and disease is, of course, complex. Densely populated and industrialized places expose their residents to a variety toxins but they usually also have excellent medical centers, emergency units, and strong scientific communities. The relationship in wildlife between population density and disease is more predictable than it is in human populations.

The world population increases by about 95 million people per year equaling three new people per second. This growth rate cannot sustain itself and at some point the Earth will not have enough resources for the human population. Environmental toxins will likely play a role in determining the Earth's *carrying capacity*—that is, the maximum amount of people that an area can support over a given time period. (It is likely that as humans reach carrying capacity, many plant and animal species will be lost to extinction.) Toxins will contribute to determining the carrying capacity in two ways: (1) shortened life span and (2) decreased fertility and potency.

Toxins shorten life span by increasing the risk of diseases such as cancer. U.S. health agencies did not record cancer statistics before 1973, but since that year the U.S. National Cancer Institute reported that cancer rates per 100,000 people rose until 1992 and have gradually declined since then. The U.S. population now has the advantage of early detection of many cancers and improved surgery and therapies.

World cancer statistics can be difficult to compile. Doctors may not know if a cancer has been caused by toxins, infections, genetic factors, or other factors. The task of studying cancer and its causes is sometimes made easier by examining trends within subpopulations, grouped by age, gender, race, occupation, and so on. In this way the medical community can uncover cancer trends and identify groups that may be at a higher risk for certain cancers than other groups.

POPULATIONS AND DEMOGRAPHICS

Environmental medicine has improved its ability to determine the risk of disease from exposure to toxins. For example health professionals have for many years warned the public of greater risks of reduced life span if a person smokes cigarettes. Environmental medicine's challenge is to predict with the same confidence illnesses due to exposure to metals, plasticizers, endocrine disruptors, or any other category of toxins. Epidemiologists draw on statistics and probability to predict the likelihood of environmental diseases.

The characteristics of a large population of individuals can be studied by dividing the population into subpopulations such as age or gender. Demographics is the study of population by breaking the population into smaller groups with related characteristics. The following example provides a comparison of two populations having different demographics that impact their health risks. The first population resides on the gulf coast of Florida and contains a large number of retirees; the average age of the community may be 55 years. The second population resides in an Oklahoma town that is home to a large university; the average age of the residents is 22 years. The two communities offer an example of age demographics. It takes little imagination to predict that the average health risks of the older community are greater than the health risks in the college town.

The following categories are the most commonly used in demographic studies on human disease:

- age
- gender
- race
- ethnicity
- country
- income
- socioeconomic standing
- education
- household size
- type of housing
- urban versus rural setting
- urban population size
- preexisting diseases

- family history of disease
- weight, blood pressure
- smoker versus nonsmoker

For each demographic category scientists collect all the data available on the causes of death or illness related to environmental toxins. The compiled data then enable scientists to calculate the incidence of a specific illness in each demographic group. From the example mentioned here, 26 out of 100,000 retirees might develop high blood pressure after exposure to toxin X, but only three out of 100,000 college students develop high blood pressure after the same exposure. Demographic categories help identify

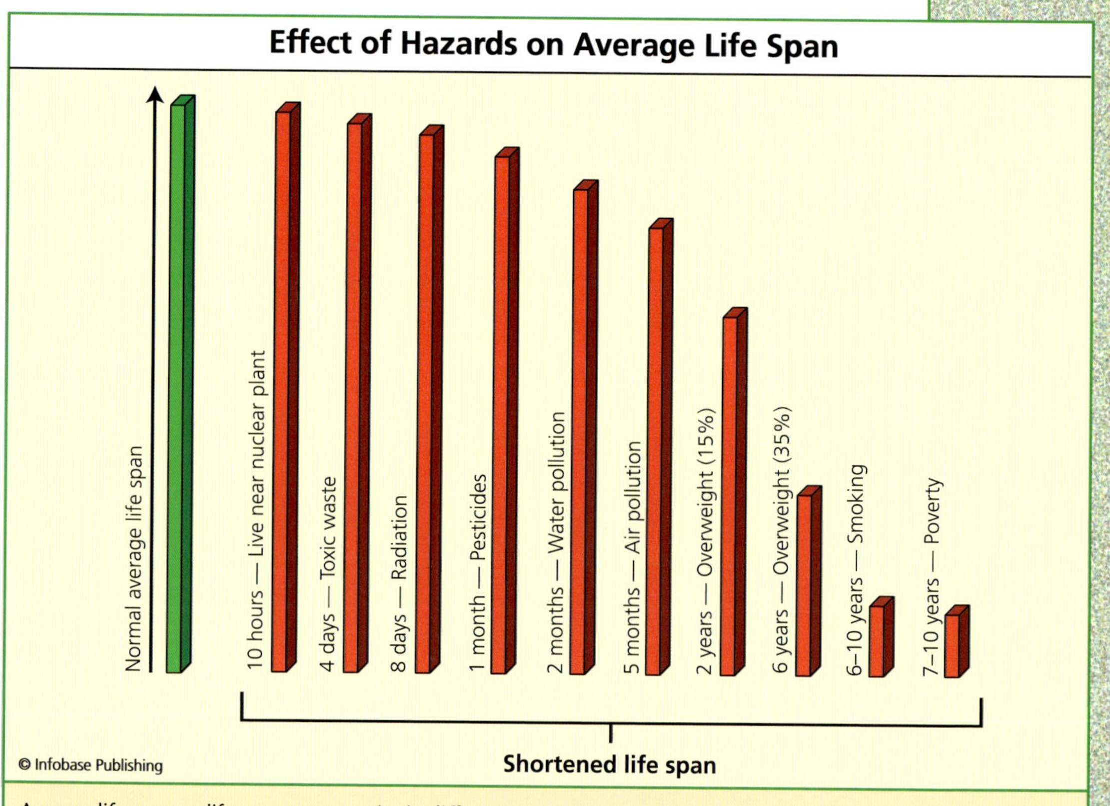

Average life span, or life expectancy, varies in different parts of the world. Canada, Australia, New Zealand, and parts of Europe have the longest average life span, more than 80 years; parts of southern Africa have an average life span of less than 40 years. Various environmental hazards shorten life span by an average of a few months to several years. In general, environmental toxins shorten life span relatively little when compared to other human-caused hazards such as smoking.

Case Study: Linking Life Expectancy to Environment

The Centers for Disease Control and Prevention (CDC) publish tables that show the life expectancies of U.S. residents since 1900. In 1900 people of all races and both genders had a life expectancy of about 47 years. To date, the life expectancy has risen to 77 years.

Linking life expectancy to conditions in the environment involves many different factors. The complex problem cannot be answered with a single statement, especially since more information must be gathered on thousands of compounds in the environment. A person can nevertheless assume that life expectancy decreases in a population with increasing levels of pollution.

The European Environment Agency (EEA) in 2007 published a report on the public health-environment connection, *Europe's Environment: The Fourth Assessment*. Europe's demographics include dense population centers, good healthcare, adequate food and water supply, and an educated populace. Europe also contains a large amount of industry. The report listed Europe's four major environmental health concerns as poor air quality, poor water quality, chemicals, and noise. One of the report's key findings was the following: "Air pollution, mainly by fine particles and ground-level ozone, continues to pose a significant threat to human health: it shortens average life expectancy in [western and central Europe] by almost one year and threatens the healthy development of children." That prediction fortunately does not take into account the roles of technology, government policy, and changes in human behavior to slow the release of toxins into the environment. Knowledge about toxins in the environment—type of chemical, amounts, persistence, health effects, and dangerous dose—still contains many gaps. Therefore, the life expectancy–environment link remains one of science's toughest problems to solve.

Scientists cannot run experiments directly on humans to determine the effects of environmental toxins, so they study trends in environmental disease and in demographic groups. These types of studies may leave room for criticism because the results may be imprecise; too many things

(continues)

(continued)

must be assumed in order to draw a conclusion. The EEA's director Jacqueline McGlade pointed out at the time of the European report, "We need to further strengthen the will to act on environmental issues across the pan-European region. This requires a better understanding of the problems we face, their nature, and distribution across societies and generations. Analysis, assessment, communication and education will help overcome this 'information gap' and will better equip those who need to act." McGlade's statement is an accurate assessment of the challenge of linking toxins and life expectancy.

the at-risk groups in any population. The connection between demographics and life span is discussed in the sidebar "Case Study: Linking Life Expectancy to Environment" on page 153.

AGE AND HEALTH

In 2004 a British poll reported by *BBC News* presented the following breakdown of respondents' greatest worry about getting old: 55 percent responded health was their main worry; 20 percent cited money; 9 percent mentioned loneliness; 6 percent worried about too much extra time; and 4 percent feared age discrimination. The Buck Institute for Age Research in California has stated that "even though people are living longer than ever, diseases of aging continue to affect many older men and women, seriously compromising the quality of their lives and their economic well-being. Cancer, in nearly all of its forms, is directly linked to aging; 77 percent of all malignancies are diagnosed after the age of fifty-five." The older demographic group therefore possesses an increased risk of harm from environmental toxins.

Aging causes hundreds of changes in the body. Glutathione made by the liver provides an example of the connection between aging and damage from environmental toxins. Glutathione is a compound from the liver made of three amino acids that detoxifies chemicals because it is an antioxidant. Antioxidants in nature detoxify many toxic compounds by adding oxygen to the compound's structure. As people age they may lose this important component. Physician James N. Balch and clinical nutritionist Phyllis A. Balch wrote in their 2000 book *Prescription for Nutritional Healing,* "As we age, glutathione levels decline, although it is not known whether this is because we use it more rapidly or produce less of it to begin with." Decreased glutathione levels may leave older people more vulnerable to illness from environmental toxins.

LIVING NEAR TOXIC SITES

Living near a toxic dump site presents a health hazard for young and old alike. The health hazard of the site may be defined by assessing three things: (1) the toxicity of the chemicals; (2) the dose of the toxins that people receive; and (3) the exposure time to the toxins. A hazardous-waste site requires a risk assessment to ensure that people understand the potential consequences of living near high levels of toxins.

When a toxic site has been discovered, scientists conduct a risk analysis to build a store of information on the environment surrounding the site. Risk assessment comes first, followed by an estimated ranking of all the risks that the site potentially holds. For instance, mercury-contaminated groundwater holds a greater health risk than surface soils contaminated with alcohol, which evaporates. The EPA also employs a Hazard Ranking System (HRS) that assigns a numerical score from 0 to 100 to each toxic site. The agency considers any site receiving a score greater than 28.5 to be a seriously contaminated hazard.

A scientific team works with EPA scientists after an HRS has been assigned to build a complete picture of all the health risks associated with living next to the toxic site. The same process takes place to assess the risks to animal health and uncontaminated natural resources such as a nearby river. Finally, the EPA notifies the public about the estimated risks of living near the toxic site.

Data have accumulated over many years on the chemicals that are known today to be hazardous. In the 1950s few people understood the extreme risk that the metal lead posed to human and animal health, but decades of epidemiological studies showed that lead poisoning has caused learning and behavior disorders in generations of children. Unfortunately, nonscientists sometimes make mistakes when assessing risk. For example,

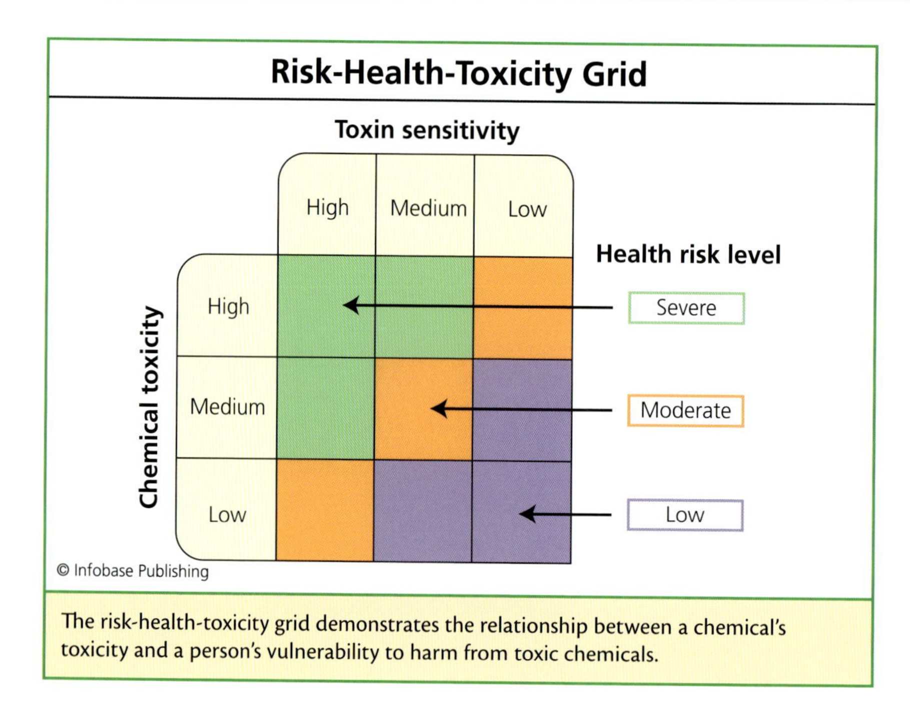

The risk-health-toxicity grid demonstrates the relationship between a chemical's toxicity and a person's vulnerability to harm from toxic chemicals.

a clean and manicured city park containing a water fountain supplying lead-contaminated water may be perceived as less of a threat than a large factory, even though the factory follows all U.S. laws on removing toxins from its emissions.

The public can make its best assessments about risk by becoming informed about environmental toxins. The following suggestions may help a family decide if the risks of living next to a toxic site are acceptable or unacceptable:

1. Understand the chemicals that potentially contaminate the site.
2. Avoid depending on only one source of information, but use a combination of resources: news articles, Web sites, university study findings, doctors' opinions, and opinions from others in the community.
3. Learn about the plans for the site. Is cleanup to begin in the near future or is the site in a long-term planning for cleanup?

4. Determine if the household has people who are at-risk individuals, such as an elderly person, a person with cancer, or a young child.
5. Assess the family's history of cancers, heart disease, or other serious health conditions.

Even without a background in science, a person can do a good job in determining an acceptable degree of risk by considering these steps. Some situations need little analysis to know that health risks are higher than the risks in the general population. The "Hazardous Site Workers" sidebar describes the extra precautions involved in known high-risk activities.

Hazardous Site Workers

Occupational medicine deals with the ailments that employees experience solely because of their job. Workers with occupations that put them near hazardous materials (abbreviated *hazmat*) get all the same illnesses that occur in the general population, but because their exposure is greater and the dose potentially higher, hazmat workers have greater health risks than other people.

Hazmat workers have a high risk of exposure to toxic substances by inhalation, ingestion, skin absorption, or injection, or more than one of these routes at once. Hazardous materials contain the following risks: explosion, ignition, reaction with other chemicals, corrosion of their storage containers, or toxicity. The activities of hazmat workers are therefore determined by several federal and state agencies, primarily the Occupational Safety and Health Administration (OSHA) and the EPA, to protect worker safety.

Worker safety depends on protective clothing that prevents the inhalation, ingestion, skin contact, or injection of a hazard, described in the following table:

(continues)

(continued)

HAZMAT PROTECTIVE CLOTHING LEVELS	
SAFETY LEVEL	DESCRIPTION
A	fully encapsulating suit, chemical-resistant gloves, and full-face self-contained breathing apparatus to protect against chemical vapors, gases, mists, and particles
B	chemical-resistant suit with secure closures at wrists, ankles, waist, facepiece, and hood, chemical-specific gloves and boots, and breathing apparatus outside the suit to protect against splashed liquids but not gases or particles
C	level B suit with the substitution of breathing apparatus that purifies the air to protect against liquids and aerosols (tiny moisture particles)
D	coveralls and chemical-resistant boots if worker is not at risk of direct chemical exposure

Hazmat suits consist of disposable synthetic material that resists damage from caustic and corrosive materials. Hazmat suits also may be covered by a metallic suit called a flash suit that protects the worker against the risk of fire or heat from an explosion. In addition to protective clothing, hazmat workers receive detailed training on hazardous wastes, proper handling of these wastes, and emergency procedures in case of accidents.

ENVIRONMENTAL DISASTERS

Accidents that release toxic substances into the environment require quick responses that identify the hazards, determine the extent of the contamination, initiate cleanup procedures, and warn the public of the hazard. Large environmental disasters usually need all of these things to happen very quickly, but in such situations communication and coordination often breaks down. Hurricane Katrina, which hit the southeastern United States in 2005, demonstrated the difficulties of communication and coordination during a massive disaster. Most U.S. communities nevertheless have an emergency response plan in case of disaster, including steps to contain toxic chemical spills.

Environmental disasters may be one of the following examples: an oil tanker spill; an oil refinery explosion; a fire at a chemical manufacturing plant; a sudden and major leak from a hazardous-waste storage site. At minimum, an emergency response for these and other disasters focuses on saving human life, minimizing environmental damage, and cleaning up the contamination. Emergency response plans typically contain the following components:

- pre-emergency planning
- roles, lines of authority, and response team communication plans
- community emergency alert system
- evacuation routes and procedures
- identification of refuge sites
- emergency medical treatment and first aid
- emergency equipment
- decontamination and cleanup procedures

The type of disaster determines the type of emergency equipment and decontamination/cleanup procedures. Otherwise, emergency response plans should fit a diversity of disasters. The best plans also contain a follow-up evaluation to identify any weaknesses or errors in the response. Large environmental disasters in the United States need action from local, state, and federal response teams. Local police departments include members that make up an emergency response team; state response often comes from the National Guard under orders from the state governor. If a governor believes that the combined local and state disaster relief is insufficient, the state may request the federal government to send help from the Federal Emergency Management Agency (FEMA). FEMA then puts its own emergency response plan into action and sends teams to the disaster site to assess the situation and provide aid. Though Hurricane Katrina exposed

The 1986 Chernobyl radioactive accident resulted in at least 57 deaths immediately or in the days following the incident. The WHO estimated an additional 4,000 people will have died in later years from radioactivity-induced illnesses. The disaster demonstrated the hazards of poor power-plant management, inadequate safety precautions, and a failure by the Ukraine government to react adequately to the disaster. Chernobyl is now deserted, and 220 tons (200 metric tons) of highly radioactive material still fill the faulty plant. *(Elena Filatova)*

flaws in the federal emergency response, the United States provides better responses than many other countries. The "Chernobyl" sidebar describes an event in history in which emergency assessment and response were insufficient during a devastating environmental disaster.

PREVENTIVE MEDICINE FOR AT-RISK PEOPLE

Health care professionals must take extra precautions in treating people with existing health conditions. This is because at-risk groups with an existing illness or injury are more vulnerable to getting a new illness. All people and animals improve their general health with a balanced energy-sufficient

Chernobyl

Chernobyl is a town in the Ukraine of the former Soviet Union where a nuclear power plant caught fire and exploded in 1986. Radiation escaped from the plant at the time of the accident and spread as far as the United Kingdom and Scandinavia. The major health threat resulted from radioactive *fallout*, which is radioactive dust and particles that drop from the atmosphere following a massive release from a nuclear facility or nuclear weapon detonation. Chernobyl's accident released mainly radioactive xenon gas, iodine, and cesium in the first few hours. Iodine 131, with a half-life of eight days, contributed to the early fatalities, and cesium 137, with a half-life of 30 years, caused longer-lasting exposure.

Chernobyl's plant exploded because of a series of lapses that allowed high-energy, high-heat reactions to go out of control. The technicians who rushed toward the fire to help put it out lost their lives to radiation poisoning. More than 30 people died within the first few minutes of the explosion, and of 61,000 Russian emergency workers, 5,000 had died of radiation-related illnesses by 1998 when death rates finally began to decrease. The 600,000 people living in the plant's vicinity will likely suffer high cancer rates and other diseases for many years.

Radioactive particles travel through the air and sometimes reach high into the atmosphere, then fall to earth for several months after the initial release. Fallout lands on soils, agricultural fields, meat- and milk-producing livestock, and surface waters that supply drinking water and fishing grounds. This created a particular hazard for the people of Chernobyl who took in radioactive doses by inhaling it, eating it, and drinking it. Government scientists from the Ukraine and international agencies have predicted that the residents of the affected area will experience a higher risk of leukemia in the future.

Two decades after the Chernobyl accident, the WHO completed a large study on the resulting deaths and illnesses. In its report, *Health Effects of the Chernobyl Accident and Special Health Care Programmes*, the WHO cited several diseases that had risen in Ukraine, the Russian Federation, and Belarus, which received most of Chernobyl's fallout. The study

(continues)

(continued)

found the following diseases to have elevated incidence rates in these countries in the 15 years following the explosion:

- thyroid disease—4,837 cases
- leukemia—doubled incidence rates to 11 cases/100,000 people
- breast cancer—slightly elevated incidence in limited studies
- cataracts—elevated incidence in small studies, especially in children
- cardiovascular disease—1,728 of the 5,000 emergency workers who died of radiation-related illness, a much higher rate than the normal population
- immunological illnesses—few data in this area
- reproductive disorders—possible increase in spontaneous abortions and stillbirths, increasing birth defects, and high rates of Down's syndrome one year after the accident
- mental, psychological, or nervous system disorders—the largest public health problem in the affected area

More than 20 years since the Chernobyl explosion, psychological problems, thyroid cancer, and breast cancer have plagued the population. The power plant today is sealed with concrete and few people live in the evacuated area, which still contains high levels of radiation.

Chernobyl's disaster occurred because of careless operations inside the plant and a lack of training among the staff. In other words, the Chernobyl disaster was preventable. The local

diet, available clean water, adequate rest, exercise, and minimal stress. Healthy individuals rely on these five factors to maintain good health. At-risk individuals, by contrast, use these factors to help the body combat disease or heal injury. In a general sense, at-risk groups possess fewer reserves to depend upon if assaulted by a second or even a third additional ailment. Environmental medicine therefore becomes critical for at-risk groups that are threatened by environmental toxins.

Impaired nutrient use or immune response puts healthy people into a risk situation. Pregnant women, for example, need adequate nutrients and

and state government made matters worse by delaying for more than a week before ordering a mass evacuation of the residents. In addition, no formal evacuation plan existed so a disorganized escape probably led to many people being needlessly exposed to fallout. The authorities initially evacuated a small four-square-mile (10 km^2) area, and took two to three years to widen the resident-exclusion zone to 1,660 square miles (4,300 km^2). By this time many more people had received some level of contamination; thousands of people have returned to the restricted zone since then. The government's ineffectiveness in managing the population in the disaster's region certainly has led to needless illness and deaths.

Nuclear industries in the United States and other countries have learned valuable lessons from the errors made at Chernobyl. A United Nations scientific commission has studied the hazards that came from the Chernobyl accident and has ongoing responsibility for compiling information on radiation exposures worldwide from any source. In the United States the Nuclear Regulatory Commission has control over the operations, training, and security at nuclear reactor sites. National nuclear power industries have shared information regarding technologies for safer power plants. Since Chernobyl, and because of the mistakes that led to its explosion and the deadly aftermath, three agencies now take the lead in monitoring global nuclear operations. The World Association of Nuclear Operators provides a forum for plant managers to share information and visit other plants as a cross-training exercise. The International Atomic Energy Agency reviews safety plans for existing power plants and new plants still in the planning stage. Organizations in the former Soviet Eastern Bloc countries such as the Ukraine have developed a hierarchy of responsibilities for safety issues, training, emergencies, and new technologies.

energy for the developing fetus and for the mother. Nutrient imbalances caused by things such as metal toxicities affect absorption and metabolism of other nutrients, and by doing so threaten pregnant women or any individual who has a higher need for nutrients. People with weakened immune systems also live at a higher than normal risk of infection from microbes and parasites. In some cases, environmental cancers attack persons already trying to fight cancer or diseases of the immune system.

The best prevention from environmental toxins is to remove the source of the toxin or, if this is not possible, move an at-risk person to a new place away from the toxin. People with higher than normal health risk should not live near known toxic waste sites. If possible, they should also avoid areas with bad air pollution and food and water suspected of containing contaminants. These actions are not always easy, especially because pollution in air and water rises and falls. Households with at-risk individuals can take some precautions, however, that provide a measure of safety. The following precautions are recommended by the National Institutes of Health, National Environmental Health Association, and the EPA for people with existing health conditions:

- Buy only certified organic foods.
- Avoid foods that are known to be associated with high toxin levels, such as swordfish contaminated with mercury.
- Drink water treated with an in-home filtration system that removes microbes and arsenic (filtration pitchers and faucet-mounted filters also provide protection).
- Wear a mask or stay indoors on days with air quality warnings.
- Avoid jobs or hobbies that require exposure to toxic chemicals.
- Practice good personal hygiene—washing hands, showering—and avoid touching hands to the eyes, nose, or mouth.
- Avoid swimming in polluted recreational waters or poorly maintained pools and hot tubs.

Even following these precautions, warding off disease cannot be 100-percent effective. Family histories and a person's genetic makeup may determine the chances of becoming ill from an environmental toxin regardless of the safety steps the person follows. All individuals must observe their environment, assess the health risks, and make wise decisions to prevent exposure to toxins.

CONCLUSION

Toxins are not toxic unless they enter an organism's system. The best defenses against taking environmental toxins into the body involve two precautions. First, all people should avoid areas, foods, and water that are likely to contain toxic chemicals or infectious organisms. Second, people can help defend against environmental toxins by maintaining good health. But subpopulations within the larger general population already suffer from health problems that make this second precaution difficult. This subpopulation contains at-risk or high-risk health groups. People with a preexisting illness, the elderly, the very young, pregnant women, and anyone with a compromised immune system make up this subpopulation. Environmental toxins present higher dangers to this group of people than the general population. For this reason the environmental medicine profession must take special care to monitor and treat at-risk individuals.

Linking an environmental toxin to any illness is one of the most difficult challenges in environmental science. Epidemiologists study groups in addition to at-risk groups to find connections between toxin exposure and a resulting disease. These studies on the world's subpopulations are called demographics. Demographics have provided much of the useful information scientists now have on environment, disease, and the risk of disease.

Demographic studies have been responsible for identifying those groups within the general population that are at above-normal risk for illness from toxin exposure. These studies have organized large amounts of data accumulated over many years to identify the major at-risk health groups and the environmental illnesses most likely to affect each of these groups. Once at-risk health groups have been determined, doctors have a much better chance of devising preventions and treatments.

Despite extensive planning to help people at higher-than-normal risk of environmental disease, sometimes the best lessons come from environmental accidents. A hazardous waste spill or a nuclear accident suddenly shows that far more people may be living in high-risk conditions than originally thought. A community living in the shadow of a hazardous-waste dump site clearly requires special safety precautions. Environmental medicine now realizes that these communities should have strict safety alerts, evacuation plans, and good distribution of information to the public about the hazards in its environment. Disaster such as the nuclear accident and release that occurred nearly 25 years ago in Chernobyl, Ukraine, have provided sober reminders of what happens when good planning and good training is absent.

The connections between the causes and health effects in environmental medicine are more difficult to identify than cause-and-effects in many other fields of medicine. Environmental medicine will always take advantage of the best new technologies for diagnosing and treating illness. It also requires a strong dependence on epidemiology methods. Part of the epidemiology of environmental disease resides in the special concerns in treating people who have an above-normal risk of injury from environmental toxins. For this reason the medical care of at-risk groups requires special, individualized attention.

7 Veterinary Environmental Medicine

Veterinary environmental medicine covers the injuries and toxicities that befall wildlife due to hazards in their environment. This is a new branch of veterinary medicine, a profession that for many years focused mainly on pets, animals of commercial value, and zoo animals. Today, veterinary environmental medicine explores a wider range of topics. Veterinarians specialize in wildlife medicine, rehabilitation methods, conservation biology, and the effects of pollution on wildlife. Knowledge on the effects of pollution on wildlife health serves two purposes: (1) it helps scientists understand the threats to species that are declining in numbers, and (2) it provides a warning of environmental toxins that may affect human health.

Wildlife becomes exposed to environmental toxins as people do, but these animals have the disadvantages of lacking medical clinics and an inability to tell a veterinarian what is ailing them. Wildlife healthcare facilities have been growing in number, however, giving many species a better chance of survival after injury or poisoning. Only richer countries of the world offer this luxury, unfortunately; most developing countries have few resources for human health so they minimize resources spent on animal health.

Wildlife rehabilitation began in the 1950s when a few small facilities offered cages and a staff to rescue small wildlife such as raccoons, foxes, opossums, birds—that veterinary clinics could not take in. The veterinary profession in the United States grew quickly between the 1960s and the 1980s and with that growth more graduates turned their interest toward

Wildlife is often dependent on places such as this one in Guatemala for water or food. Environmental toxins and wastes in such areas have had unknown but probably very serious effects on wildlife survival. This river contains plastic bottles, tarpaulins, detergents, and sewage.

wildlife care. Wildlife veterinarians stationed in rehabilitation centers now specialize in an array of animals from reptiles to coyotes. Larger animals such as mountain lions, moose, or bears usually become the province of zoos or rehabilitation centers specializing in those species.

This chapter reviews the current activities in wildlife medicine. The chapter covers medical threats to wildlife from human-caused pollution, as well as zoo animals, capture-and-release techniques, and the expanding area of conservation biology. The chapter then closes with a section on veterinary medicine's advances in wildlife health and ideas for the future.

WILDLIFE MEDICINE

Wildlife encompasses any nondomesticated animal, even if the animal had been bred, born, or hatched in captivity. Wildlife families consist of the following groups of species: reptiles, amphibians, fish, birds, terrestrial mammals, marine mammals, and aquatic wildlife. At present many species within these groups do not seem wild at all because their numbers have declined to a point at which they exist only in protected habitats. For example, few American bison roam freely over the western plains; most bison have been confined inside park boundaries and private ranches. American bison nevertheless are wildlife.

Wildlife medicine contains five main research areas: (1) rehabilitation; (2) ecology of diseases; (3) conservation medicine; (4) wildlife–domesticated animal relationships; and (5) clinical advances. Rehabilitation consists of medical activities, starting from rescuing an injured or diseased animal to providing medical care, giving the animal a safe environment for recovery, and then assessing whether a rehabilitated animal can or cannot return to the wild. The ecology of disease comprises the science of new diseases in wildlife populations from either infections or toxic chemicals. New diseases, called emerging diseases, arise and spread throughout a wildlife community just as they do in a human population. Conservation medicine focuses on the care of threatened or endangered species.

Veterinary medicine has made important advances in the diagnosis and treatment of wild animals. In many cases, animals receive state-of-the-art care. This veterinarian in Delhi, India, is scanning a microchip that was implanted just under this elephant's skin. *(Wildlife S.O.S., PerLvr.com)*

The two newest areas in wildlife medicine are wildlife–domesticated animal relationships and clinical advances. Wildlife–domesticated animal relationships include the transmission of disease from domestic to wild animals or from the wild to domestic animals. Unvaccinated pets have a high risk of contracting distemper, for example, and they can then spread the disease to raccoons, opossums, coyotes, and other wildlife. Often a disease outbreak will decimate local wildlife species. Conversely animal owners often fear wildlife diseases that potentially threaten the health of domesticated animals. American bison provide one such example;

ranchers believe bison will infect their cattle with contagious brucellosis, so they have shot and killed any bison that wander in the direction of ranchland. The second new area, clinical advances in wildlife medicine, consists of groundbreaking developments in surgery, trauma care, anesthesia, drug therapy, prostheses, and vaccinations. Clinical advances have had an impact in saving terrestrial species, marine mammals, birds, and zoo and aquarium animals.

Veterinarians in wildlife medicine also participate in wildlife health policies, conservation policies, and habitat monitoring. Wildlife medicine has an overall effect on the health of people and domesticated animals, but perhaps most important, monitoring wildlife provides a picture of the threats from pollution and habitat loss that wildlife face.

URBANIZATION AND FARMING

Loss of habitat creates the primary reason that plant and animal diversity worldwide has declined. Some animals adapt to life in places altered by humans better than other wildlife. Wildlife that cannot adapt must find new habitat that can sustain its populations; if these animals cannot do this, they head toward extinction. Adaptive species, by contrast, find ways to thrive in cities even though these species did not evolve for this possibility.

Dense urban areas put unique hazards in the way of animals, perhaps best demonstrated by the plight of many birds living in or near urban centers. Since the 1960s, some bird population numbers have declined 90 percent, chiefly because of habitat destruction. But the other leading causes of bird deaths, listed here, illustrate an abundance of threats from urbanization:

- invasive plants that outgrow native seed-producers
- wind-power turbines in migration routes
- collisions with lighted and glass encased buildings and windows
- lighted cellular phone towers
- vehicles
- domestic and feral cats

- pesticides
- disease
- climate change

Each of the above factors arises partly or wholly from human activities in urban settings. Even adaptive species find it difficult to adapt to the rapid pace of new technologies. Karen Imparato Cotton is a bird crash specialist at the American Bird Conservancy. In 2008 she described to the *San Francisco Chronicle* the effect of brightly lit tall buildings (light pollution) on bird migration. Many bird migrations take place at night, and Cotton pointed out, "The light fields entrap night-migrating birds. They seem reluctant to leave these lit areas and tend to circle within them. As they pile up in the light field, circling the structure, they collide with each other, with the building, or they collapse from exhaustion." Lights, noise, and the constant presence of humans cause similar harmful effects to migrating mammals and aquatic life.

Far from urban areas, grasslands have become one of the most threatened *biomes* on Earth. The loss of grassland habitat affects birds, ground-dwelling mammals, small predator mammals, reptiles, and soil invertebrates. Grasslands in the United States that have previously been protected by farmers return to cultivation for at least one reason: rising feed prices. Cattle operations find it cheaper to cultivate their own feed than to buy it when feed prices increase as food demand increases. This occurrence may continue as the world's human population grows. The industrialized world also demands alternative fuels made from crops instead of petroleum. These so-called *biofuels* leave less feed available for food animal and crop production, and so prices rise.

Farming operations present grassland animals with additional health threats beyond habitat destruction: injury or death caused by farm machinery; losses to dogs; pesticide exposure; loss of prey; and hunting.

DISEASE INDICATORS

Animals in the wild can serve as warning systems of the spread of environmental toxins. Perhaps the most famous example of wildlife harm alerting the public to an environmental toxin came from the effect of the pesticide dichlorodiphenyltrichloroethane (DDT) on bald eagles in North

America from the 1940s until 1972, when the United States banned its further use. DDT damaged bald eagle reproduction by interfering with eggshell formation and caused eagle numbers to plummet. The dramatic effect of DDT on wildlife health signaled the problem of environmental toxins and their threat to human health.

Wildlife medicine makes use of two different roles that animals play in the environment: indicators and sentinels. *Indicator species* give early warning of ecosystem destruction. Specific bird, insect, and marine organisms provide information about their habitat or ecosystem merely by being present or absent in the habitat. For example, marine nudibranches—nicknamed sea slugs despite their brilliant colors, which are absent on terrestrial slugs—serve as an indicator species of healthy marine habitat. The presence of diverse nudibranches in an area indicates that corals, sponges, and marine invertebrates are probably thriving. A lack of nudibranches or very poor nudibranch diversity indicates a degraded habitat.

Disease indicators alert scientists to the degradation of habitat specifically because of environmental toxins, chemical or biological. One example of an environmental toxin indicator involves the use of fish to warn of acid runoff. Acidic water injures gills, eyes, and skin of many fish species, and by selecting one or two species as indicators, scientists can monitor acid rain pollution of habitat.

The behavior of two special indicator species provides information on ecosystem and habitat, respectively: *keystone species* and *foundation species.* Keystone species serve ecosystems because many other species in the same ecosystem depend on the keystone for survival. A foundation species, by contrast, enhances habitat by its behavior, such as digging burrows, rooting through soil, clearing out dense plant growth, or hollowing out dead trees.

While indicator species tell a story about habitat, *sentinel species* provide information about potential threats to human health. Environmental medicine makes use of sentinel species to monitor new toxins emerging in food, water, or air. The sentinel animal's heart rate, respiratory rate, reproduction, and blood and urine chemistry are all vital signs that may give clues of toxin exposure.

Taken together, indicators and sentinels give an overall picture of the threats to wildlife health. Veterinarians and rehabilitation centers study indicator and sentinel species to discover clues about the illnesses they see in their patients.

ENVIRONMENTAL TOXICITIES IN ANIMALS

Like humans, wild mammals ingest or inhale toxins or absorb them through their skin. After the toxin has entered the mammal's body, it travels in the bloodstream and affects organs and tissues in a similar way as it does in people. Toxins' effects on wildlife can be lethal or sublethal (dangerous but not deadly), acute or chronic, and toxins exert either indirect effects or direct effects in an individual, or both. Some toxins, however, affect certain species and not others. For instance, insecticides harm fish and mammals more than fungicides do.

Heavy metals poison freshwater and marine fish, reptiles and amphibians, waterfowl and forest birds, and insects, as well as mammals. Heavy metal ingestion by wildlife has caused reduced fertility, birth defects, retarded growth, anemia, cancers, poor immune responses, and neurological damage. Some metals cause the most harm as pure elements, and other metals, such as mercury in the form of methylmercury, cause increased toxicity when part of an organic compound. Wildlife suffers risks of contamination from all heavy metals, but mercury and lead have produced the most dramatic effects, so wildlife studies often take place on these two metals.

Because fish live at the bottom levels of many food chains, they serve as an entry point for mercury in food webs and ecosystems. Animals higher up the chain then bioaccumulate mercury. The National Wildlife Federation's 2006 report, *Poisoning Wildlife: The Reality of Mercury Pollution*, described the pervasiveness of mercury in animal populations today: "The conventional thinking was that because mercury can easily be converted to methylmercury (the toxic form that accumulates in living things) in water, only species that live or feed in aquatic habitats are at risk of exposure." But environmental studies showed that the mercury did not stay confined to water. The report said, "A recent example on a river stretch in Massachusetts demonstrated that wetland birds, such as the Red-winged Blackbird, have average mercury burdens that are five times higher than associated fish-eating birds such as the Belted Kingfisher. Research is also showing high mercury levels in songbirds. . . . indicating that our forest habitats are accumulating harmful levels of mercury as well." Predators of small birds like foxes, raccoons, hawks, and owls accumulate greater mercury levels in their bodies, and so on up the food chain. The National Wildlife Federation determined that the following aquatic or terrestrial animals are at the greatest risk of mercury toxicity, or already exhibit signs of toxicity:

- fish—walleye, largemouth bass, brook trout, northern pike, yellow perch
- mammals—river otter, mink, raccoon, Florida panther, Indiana bat
- fish-eating birds—bald eagle, great egret, loon, kingfisher, tern, wood stork, diving ducks
- songbirds—thrushes, warblers, vireo, Carolina wren
- reptile—American alligator
- amphibian—bullfrog
- invertebrate—crayfish

Scavengers, birds, and animals receive toxic levels of lead when they ingest lead bullet fragments off the ground. Inside the animal's body, lead crosses cell membranes and binds proteins inside the cell. In warm-blooded animals, lead blocks hemoglobin's ability to carry oxygen in red blood cells, leading to death in animals and humans by suffocation. Animals such as vultures, condors, crows, coyotes, wolves, small mammals, and deer can suffer lead poisoning from just one or two ingested bullet fragments. Similarly, waterfowl such as loons that dive under the surface to feed swallow lead sinkers, which can be fatal.

The University of Minnesota Raptor Center has described lead poisoning in birds as follows: "A bird with lead poisoning will have physical and behavioral changes, including loss of balance, gasping, tremors, and impaired ability to fly. The weakened bird is more vulnerable to predators, or it may have trouble feeding, mating, nesting, and caring for its young. It becomes emaciated and often dies within two to three weeks after eating the lead." Conscientious hunters now use non-lead ammunition, but lead in the environment still poses a danger to wildlife.

Industrial chemicals have become as pervasive in wildlife habitats as they have in human communities. The U.S. Environmental Protection Agency (EPA) has created an online database named Ecotox, which provides information on the effects of chemicals on the health of aquatic things and terrestrial animals. Ecotox information covers the following animal categories: amphibians, birds, crustaceans, fish, insects and spi-

Known Effects of Specific Chemicals on Nondomesticated Animal Species		
Reproduction	**Physiology**	**Behavior**
hormone function	cell biochemistry	social group behavior
genetics	histology	food avoidance
fetus development	intoxication	water avoidance
population	intracellular enzyme activity	lack of human avoidance
breeding behavior	extracellular enzyme activity	general behavior
growth	bioaccumulation	breeding behavior
mortality	hormone function	
morphology	immune system function	
egg strength	injury	

ders, other invertebrates, mammals, mollusks, reptiles, and worms. The EPA has further sorted the information by animal species, chemicals, and signs of toxicity, which helps wildlife caregivers diagnose toxin poisoning. Twenty-four major toxic signs have been noted in animal species caused by specific chemicals, summarized in the table above.

The health effects listed in the table may result from a specific species reacting to a specific chemical, or they may be more general effects associated with more than one toxin. Therefore, the type of animal and toxin play a role in determining the main effects of toxin poisoning. Toxins, furthermore, affect only tissues that have particular receptor sites on the outside of their cells to which the toxin can bind. In animals, the highest concentration of a chemical in an organ does not necessarily correlate to where the chemical does the most harm.

Common pesticides used today belong to three major groups: organophosphates, carbamates, or pyrethroids. Each of these categories is lethal to birds, mammals, and fish at high levels. Pesticides cause many of the same general effects on living tissue as industrial chemicals, and both cause direct and indirect effects on animals. Direct effects resemble those seen in humans: illness, reproductive disorders, changes in behavior, and death. Indirect effects may take the following forms:

- elimination of food sources such as insects and aquatic invertebrates
- stunted growth due to disrupted food chains
- decline in reproductive rates in nutrient-depleted animals
- reduction in plants that provide cover
- reduction in nest-making materials
- upset migration patterns due to behavioral changes or lack of food

Pesticide labels used in the United States must indicate any dangers to wildlife and the warnings must be stated in one of the following ways:

- "Hazardous to Humans or Animals"
- "This Product Is Toxic to Birds and Other Wildlife"
- "This Product Is Toxic to Fish"
- "This Product Is Highly Toxic to Bees"

People can help preserve the quality of wildlife habitat by taking the following eight precautions when using pesticides: (1) avoid spraying on windy days; (2) avoid washing pesticide applicator equipment in natural waters such as streams; (3) avoid using pesticides if rain is expected to wash the pesticides off; (4) avoid applying pesticides in areas where the water table is close to the ground's surface; (5) set up buffer zones between pesticide areas and wildlife habitat; (6) begin pesticide applications in the middle of cropland to allow wildlife time to escape; (7) avoid applying near bees; and (8) read and follow the directions on every pesticide product.

Oil spills have had especially dire consequences on wildlife that live in marine habitats or depend on shorelines for feeding or habitat. Oil injures the outside of animal bodies as well as inner organs when ingested. Oil ingestion poisons the milk of nursing mothers and passes toxins to nursing young, and it affects sea and coastal birds by altering feather structure, which in turn prevents the bird from maintaining body temperature. Hypothermia (lowered body temperature) soon overwhelms an oiled bird. Waterfowl additionally need feathers for buoyancy, but oil-marred feathers make birds less buoyant so they expend more energy to swim. The energy-depleted birds cannot hunt for food as efficiently as healthy birds and become even more at risk because of difficulties in taking off and flying.

Oil affects sea mammals in two main ways. First, sea otters and fur seals suffer injury from oil when the material damages their fur coats because they depend on fur for maintaining body temperature and for swimming. Second, marine mammals that depend on blubber rather than fur for warmth—sea lions, dolphins, whales—are injured mostly by oil irritation to their skin and by ingesting oil.

In 2007 an oil tanker rammed the San Francisco–Oakland Bay Bridge in San Francisco, California. The spill covered hundreds of miles of beaches and injured and killed thousands of waterfowl. This oil-covered grebe on the beach is still alive but required immediate care to survive. Wildlife care centers cleaned and saved hundreds of birds, but many more never received prompt help and died. *(Tome Prete)*

Oiled animals that receive quick care can recover from oil toxicity. Such care involves gentle washing of feathers or fur with mild detergent and frequent flushes of the digestive tract. Despite encouraging successes in rescuing oiled wildlife, oil spills remain one of environmental medicine's biggest problems.

Wildlife that manages to avoid harm from chemicals confront two additional health threats. The first is garbage left on beaches or in forests. Ingested items injure the digestive tract or block it. In some cases, animals die by starvation because an injury has prevented feeding. Discarded fishing wire often wraps around fishing birds such as pelicans, tying the bill shut. Domestic animals pose a second threat to some wildlife as a reservoir for contagious disease. Domesticated pets that have not been vaccinated against bacterial or viral diseases, for example, can infect wildlife, which have no defenses against these pathogens. Some domesticated animals have been trained to help rather than threaten wildlife, as described in the "Environmental Detection Dogs" sidebar on page 178.

CAPTIVE POPULATIONS

Zoo animal medicine is a specialty in veterinary environmental medicine that focuses on exotic animals and species housed in zoos. Zoo animal

Environmental Detection Dogs

Dogs possess an olfactory system 50 times larger than a person's and have a sense of smell that is 44 times stronger than the human's olfactory ability. For this reason dogs have been trained to detect illegal drugs, explosives, people, and fatalities. Dogs have already been put to work to detect accelerants used in arson, so there is little surprise that dogs might be used to detect toxic chemicals in the environment. As long as a chemical is not found throughout the environment, a trained chemical detection dog can locate specific chemicals. In this way chemical detection dogs may be used to find unknown waste dumps, vapor leaks from groundwaters, illegal pesticides, or toxic molds, and these dogs also signal the presence of toxic compounds in a home, either in the air, tap water, or in items such as furniture. Some dogs receive specialized training to locate banned pesticides on farmland. Others called mine detection dogs work at coal mines as early warning systems for explosive gases. At least one dog has even been trained to find tiny amounts of mercury in the environment.

Biological detection dogs play a role in wildlife conservation. These dogs have been used in helping biologists find rare species, and they also find the scat (solid waste) of predators, such as grizzly bears or mountain lions, rather than taking on the dangerous mission of tracking. Laboratory technicians then analyze the DNA in scat to determine the species, sex, individual identity, diet, health condition, stress, and how closely individuals in a population are related—relatedness often indicates if a species or a habitat is declining. It seems detection dogs' ability to analyze nature make them more valuable than even the most advanced laboratory equipment.

medicine, or *zoological medicine,* includes zoo animals' nutrition, pathology, and clinical medicine (which encompasses disease therapy and surgery). These advances have had an important positive impact on the conservation of endangered species.

The world's best zoos provide preventive care to the diverse species they hold. Preventive medicine in zoos consists of routine examinations, vaccinations, parasite testing and deworming, and observation of sick and

Zoos have had poor reputations in the past in wildlife care, but today the world's leading zoos offer excellent medical care. They also conduct studies that could save declining animal populations, such as the cheetah's, shown here in its natural habitat in Tanzania. *(Rob Qld)*

well animals to learn more about their signs of disease. Sick animals in the wild are vulnerable to predation by other species or harm from animals of their own species. In fact, a small proportion of animals in the wild die of natural causes; as they become infirm, predators kill them. For this reason animals try to hide any sign of illness for as long as possible. Practitioners of zoological medicine must be able to spot the slightest clues that an animal may be ill. In preventive medicine, doctors also quarantine new animals to avoid introducing disease into the zoo population. A quarantined animal lives apart from the general population until veterinarians confirm that the animal is healthy.

Zoological medicine has made remarkable advances in animal care and includes the veterinary specialties such as cardiology, ophthalmology, dentistry, surgery, and rare diseases. Zoological medicine employs many of the same techniques as used in human medicine: vital-sign monitoring, injections, X-rays, ultrasound, intravenous feeding, and catheterization.

In many cases zoo veterinarians see ailments in captive animals that were not previously known.

In 2008 the Smithsonian's National Zoo in Washington, D.C., nurtured a new family of rhea chicks—a rhea is a large, flightless, ostrichlike bird from South America. ScienceDaily quoted the zoo's biologist Sara Hallager in explaining the value of this event to veterinary medicine and conservation: "The birth of these chicks comes at a time when the number of rheas in zoos is declining. By developing . . . an official record relating to the animal's history, we are hoping to ensure that knowledge relating to their husbandry is retained and to stimulate further interest in the birds both inside and outside of zoos." Hallager's words may sum up the worth of holding certain rare and endangered species in captivity: By learning everything they can about these animals, veterinarians can help others in threatened or polluted habitats.

Pathology serves as an important learning tool in zoological medicine. Each zoo animal that dies in captivity today, either from old age or other conditions, should undergo an autopsy. This postmortem exterior and interior examination uncovers the cause of a disease, or at least identifies the organs affected by the disease. Animals known to have been exposed to environmental toxins provide veterinary medicine with valuable information on how these toxins injure the body. Necropsies provide information on things in addition to toxin exposure: mode of action of new infectious diseases; tissue and cell abnormalities; effects of nutritional deficiency; the effect of metabolic disorders on organs; and unknown differences between animals of a species due to age or gender.

Even with the newest technologies in treating animals, zoological medicine is a difficult undertaking. Many zoos house more than 300 diverse species with different body types and physiologies—zoos own an amazing variety of instruments for examining, treating, and restraining animals. Stressed or sick animals usually refuse food, which delays medical recoveries, and zoo animals display a range of behaviors. Some are docile; others are strong and deadly. Animals such as primates are intelligent and cunning and they can be clever about resisting examinations.

Rehabilitation centers differ from zoos because they hold animals in captivity only as long as needed to recover from toxicity or disease or to heal an injury. These centers release only animals that have made a full recovery. Rehabilitation centers differ from zoos in two additional ways: (1) rehabilitation centers house injured animals from the local community,

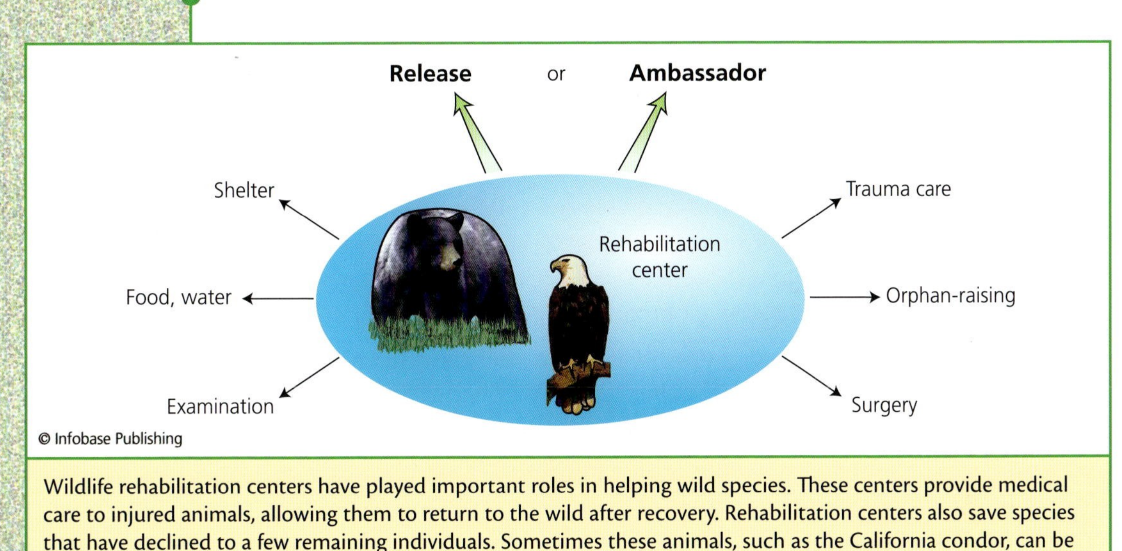

Wildlife rehabilitation centers have played important roles in helping wild species. These centers provide medical care to injured animals, allowing them to return to the wild after recovery. Rehabilitation centers also save species that have declined to a few remaining individuals. Sometimes these animals, such as the California condor, can be restored to health in captivity, then released back into the wild to reestablish a natural population.

specifically for treatment and release back into the wild, and (2) rehabilitation centers' purpose is toward wildlife treatment rather than wildlife display for the public. *Ambassador* animals provide one exception to the second role of rehabilitation centers. An animal that is too badly injured to be released following rehabilitation may stay at its rehabilitation center as an ambassador, living in a safe enclosure where the public can view it. Ambassadors in turn serve as educational tools to teach students and the public about wildlife and the threats to wildlife. Marine mammal rescue centers offer such combinations of rehabilitation and education, highlighted in the "A Day in the Life of a Rescued Harbor Seal" sidebar on page 184.

THE HUMAN AND ANIMAL HEALTH CONNECTION

Zoos and rehabilitation centers monitor the types of ailments shared by animals and people, especially diseases that humans contract from animals, called *zoonotic diseases.* Veterinarians and medical doctors know of a number of zoonotic diseases: Lyme disease from deer; hantavirus from ground squirrels; influenza from birds; and rabies from skunks, bats, foxes, and raccoons. As people move into previously undisturbed animal habitat, doctors will probably discover new zoonotic diseases.

A Day in the Life of a Rescued Harbor Seal

Marine mammal centers operate along the U.S. Pacific coast for the purpose of saving and treating animals in distress. These centers accept captured or incapacitated animals, administer treatment, and hold animals for observation. When a marine mammal shows signs of complete recovery, the center returns the animal to its natural habitat.

A typical rescue of a harbor seal in California's Monterey Bay begins with a member of the public seeing the animal stranded on a beach and calling a local rehabilitation center for help. Mammals such as seals strand on beaches (requiring *beach rescues*) or in shallow waters near beaches (requiring *water rescues*) for the following main reasons: either the animal is injured or otherwise incapacitated so it cannot return to the water, or it is a healthy animal that cannot return to the water without assistance, such as an animal tangled in fishing line or netting. In either case the animal might well need medical attention.

The person who finds the seal stays back from the animal at least 50 feet (15 m) and keeps all dogs and other people away from it. The observer describes for the rescue team the seal's condition (wounds, emaciation, an orphaned pup) and its location on the beach. The rescue team arrives on the scene as soon as possible and assesses the seal for species, size, age, physical condition, dehydration, behavior, vocalization, interaction with humans, and wounds from boats, gunshots, or sharks. A severely injured seal would likely be lethargic, nonvocal, and unable to avoid humans. The rescue team captures the seal in a net or the team may be able to herd the seal using herding boards so that the seal moves on its own power into a crate. Water rescues of entangled animals require the use of wet suits, tools to cut lines or netting, and possibly a boat to bring an injured animal to shore.

At the rehabilitation center, if the harbor seal has a serious wound, poisoning, or disease, the staff takes it to the intensive care unit (ICU). In the ICU technicians draw a blood sample, check respiratory rate, collect a feces sample, and may take a sample from the wound. Wound samples provide information on microbial or parasite infection. The seal may then be X-rayed, or examined by ultrasound or endoscopy. An animal that has been poisoned receives digestive tract

Conservation medicine focuses on the relationships between environmental toxins, wildlife and human health, and biodiversity. Global warming is expected to cause new diseases to human populations, and the same thing will likely take place in animal populations worldwide. Veterinarians in conservation medicine study newly discovered diseases in all animal species. Conservation medicine also includes the study of

flushes with water, intravenous fluids, and drugs to prevent convulsions. If the seal shows signs of cancer, it would likely be euthanized to prevent further pain. Frances Gulland, head veterinarian at California's Marine Mammal Center, told *Bay Nature* magazine in 2007, "We see cancer in about 17 percent of the adult sea lions that die at the Marine Mammal Center, and they all have high rates of PCBs and DDT in their blubber." Marine mammal centers therefore play an important role in assessing the effects of organic chemicals on marine health.

The center puts the harbor seal into a private cage that offers quiet and minimal contact with people and noise. The cage's swimming pool provides easy entry and exit and its water contains temperature gradients so the seal can move to whatever water temperature feels most natural to it. The seal receives regular feeding, filtrated drinking water to prevent infections, and round-the-clock monitoring during its time in the ICU.

A harbor seal with minor injuries, showing alertness, appetite, and an interest in the activity of other animals, will share its recovery pen with two to three other harbor seals. The pen-mates exercise in their pool and receive regular feedings of the fish they normally eat and clean drinking water. The center's staff clean and sanitize pens on a regular schedule. Recoveries may take from months to years before an animal is ready to return to its natural home. Severely injured animals may need up to two years for full recovery from injuries and to undergo reconstructive surgeries.

Release criteria include the following: stable and normal body weight and blood parameters; the ability to eat on its own; and, if it is a pup or a vision-impaired animal, it must be able to track and catch food on its own. On release day the harbor seal receives a tag on its flipper that identifies the rehabilitation center; it may also have a tiny transmitter implanted under the skin for tracking. The seal then rides caged in a vehicle to a remote beach that lacks human activity. A crew sets the cage near the water and opens the door. The healthy harbor seal then scrambles down the ramp and into the ocean to live the rest of its life in its natural habitat. Marine mammals now receive medical care that rivals the care given to terrestrial animals.

environmental toxin effects on ecosystems that have a direct effect on human health.

Conservation medicine currently supports global health issues in the following ways: (1) surveilling global disease; (2) identifying emerging disease hotspots (defined areas with a dense concentration of disease, biodiversity, endangered species, etc.); (3) tracking wildlife mortalities and causes; (4) monitoring pollution effects and dispersion in the oceans; and (5) studying the reasons for high incidences of animal deaths in aquatic or land habitats. Conservation medicine makes use of sentinel species, such as the species listed in the table on page 185, to assess the health of wildlife.

Marine mammals receive medical care and recover from injuries and disease in rehabilitation centers. Shawn Larson, Ph.D., curator of animal health and research at Seattle Aquarium, feeds a young sea otter. *(Shawn Larson)*

Sentinel Animals Used in Marine Conservation Medicine		
Sentinel Animal	**Main Location**	**Assessment**
sea turtles (loggerhead and black turtles)	Baja California	fibropapillomatosis tumors, an indicator of emerging disease, changing water temperatures, environmental toxins, or chronic stress
California sea lions	California coast	environmental toxins in blubber
brown pelicans	California coast	West Nile virus, Newcastle disease, avian influenza, algae bloom toxins
Antillean manatees	Florida, Mexico, northern Caribbean, Belize	persistent toxins in blood or blubber
La Plata dolphins	Argentina	emerging diseases, marine toxins
migratory neotropical birds and waterfowl	Mexico	West Nile virus, avian cholera

A branch of conservation medicine focuses on the effects of environmental toxins on pets. Diseases in pets that are monitored by veterinary clinics may provide insight on the health of like species in the wild. For example, information may be gained on the dangers of household chemicals by watching the unhealthy effects these products have on domesticated birds, dogs, cats, rabbits, or rodents. In 2008 the Environmental Working Group (EWG) reported that dogs it tested contained 35 different chemicals in their blood or urine of which 11 caused cancer, 32 caused reproductive disorders, and 24 were neurotoxins. In cats, of 46 chemicals, nine caused cancer, 40 damaged the reproductive system, 34 were neurotoxins, and 15 damaged the endocrine system. Altogether the EWG and

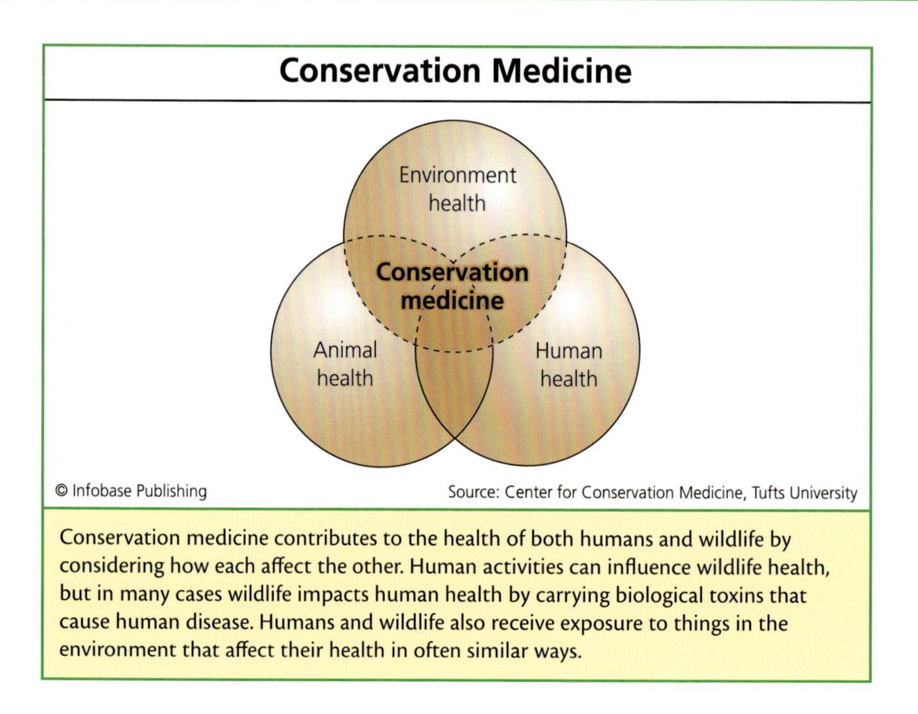

Conservation medicine contributes to the health of both humans and wildlife by considering how each affect the other. Human activities can influence wildlife health, but in many cases wildlife impacts human health by carrying biological toxins that cause human disease. Humans and wildlife also receive exposure to things in the environment that affect their health in often similar ways.

the Centers for Disease Control and Prevention (CDC) have concluded that these pets contain as much as five times the amount of mercury in their bodies as humans; cats may have as much as 23 times the amount of flame retardants. The EWG's Olga Naidenko told the *Milwaukee Journal Sentinel,* "These dogs and cats were highly polluted." Domesticated pets have become unwillingly experiments in toxicology.

Pet birds suffer extreme effects from certain household items. Fumes from cooking with nonstick coatings and secondhand cigarette smoke cause toxicities in many exotic pet birds. Lead or zinc poisoning causes nerve damage and can lead to death. California condors have become a rare bird rather than an exotic pet, but as the sidebar "Case Study: Condors Reintroduced to Baja California, Mexico" (see page 187) discusses, condors also suffer dire consequences from lead poisoning.

CLINICAL ADVANCES IN WILDLIFE HEALTH

Wildlife medical care has advanced to sophisticated methods for saving lives and repairing injuries. Veterinarians now administer superior

Case Study: Condors Reintroduced to Baja California, Mexico

Condors in the Western Hemisphere may be either Andean condors from the mountains of South America or California condors that reside in limited habitats in California and Arizona. Both of these large vulture-related birds are endangered and their numbers have come perilously close to extinction. Andean condors now number about 1,000 birds and California condors number between 100 and 200 in the wild, though their habitat once stretched from the U.S. west coast to New York. In 2005 members from the California Condor Recovery Program released a female condor in a new site, the Sierra San Pedro de Martír national park in Baja California, Mexico. Nearly all of the California condors now living in the wild had been hatched in condor centers in Los Angeles or San Diego, California, Boise, Idaho, or Portland, Oregon—the bird released in Mexico had been raised in San Diego's condor recovery center. But this condor like so many others will face daunting challenges in the wild.

All of the reasons condors have slipped toward extinction have not been uncovered. Ranchers killed some of the huge birds in the past—they have a nine-and-one-half foot (2.9 m) wingspan—and habitat loss and scarcity of prey probably added to the declines. One aspect of the condor's endangered status may be their feeding behavior: They eat only freshly killed wildlife carcasses. Hunters who use lead bullets may leave lead fragments in the carcasses after cleaning a kill. Lead poisoning damages the bird's crop, an organ in the digestive tract; the crop shuts down and the condor starves.

Three approaches to reducing lead poisoning deaths have been introduced to help condors. First, biologists put carcasses recovered from cattle farms out in the condor habitat to increase the chances that condors will ingest clean lead-free food. Second, efforts have been slow but steady in substituting non-lead ammunition for lead bullets. In 2005 environmental toxicologist Michael Fry told the *Smithsonian Zoogoer* magazine that 60 percent of condors had higher than normal lead levels and 15 percent had near fatal levels. California and Arizona have published educational articles in hunting magazines to encourage a switch to copper or

(continues)

(continued)

other non-lead bullets and shot. Jeffrey Miller of the Center for Biological Diversity was quoted in the newsletter as saying, "Voluntary measures clearly won't do it all. It doesn't make sense to release condors into the wild without addressing the lead problem." On July 1, 2008, California banned the use of lead bullets in areas where condors live. The new regulation came not a moment too soon; in early 2008 seven California condors experienced a slow and painful death due to lead poisoning. The U.S. Fish and Wildlife Service's condor expert Jesse Grantham told the Associated Press, "This is the highest lead exposure event we've had in ten years." Until the lead ban begins making a difference in condor mortalities, the birds may be dependent on the third approach to reducing deaths: chelator treatment. Chelator compounds draw lead out of the birds' tissues and make the lead easier for the body to excrete.

Two years after release, people spotted the Baja California condor near San Diego. The bird circled before turning south to Mexico, but the foray across the border gave hope that a population will establish a permanent habitat along the California and Mexico coast. More condors are being released annually in Sierra San Pedro de Martír until the park reaches its carrying capacity of 20 breeding pairs. These new releases will face the same threats as birds to the north until lead starts declining in condor habitat due to lead bans. Condor recovery groups and local governments can make an important contribution in the story of the condor's return.

care and recovery methods, medications, anesthesia, and surgical techniques. Clinical sciences use cutting-edge technology taught today at U.S. veterinary schools in the following disciplines: anesthesiology, oncology, dermatology, cardiology, radiology, neurosurgery, ophthalmology, reproduction, internal medicine, and critical care.

Significant advances in animal health relate to improved drug treatments and medical equipment, such as the following:

- ventilators for respiratory distress
- dialysis for kidney failure

- vital sign monitors for heart rate, lung activity, and blood pressure
- defibrillators for cardiopulmonary resuscitation
- diagnostic tests
- radiographs (X-rays), fluoroscopy (motion X-rays), ultrasound, computed tomography (CT scans), and magnetic resonance imaging (MRI)
- endoscopy for gastrointestinal, urinary, or respiratory tracts
- echocardiography
- electrodiagnostic testing for muscle and nerve disorders

Cancer treatment involves two kinds of technology: (1) radiation using a linear accelerator unit that delivers a dose of high-energy X-rays to a cancer site or a tumor, and (2) chemotherapy. The type of radiation or chemotherapy a veterinarian selects depends on the type of cancer diagnosed in an animal. Wildlife medicine may not have all the information it needs about certain cancers in rare wild species, so treatments must be based on the treatments used on domestic animal cancers. The table on page 190 describes the major chemotherapy treatments that have been shown to work in domesticated animals; some of these drugs have already been tried on wildlife with cancers or tumors.

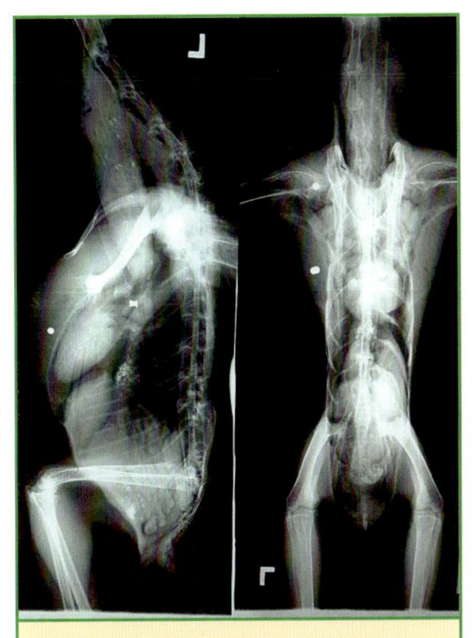

Like terrestrial species and marine mammals, birds receive care in specialized veterinary clinics. X-rays such as this taken on a great blue heron can aid in injury diagnosis. The dense (white) objects in the bird's thoracic cavity could be lead pieces. *(International Bird Rescue Research Center)*

Trauma care has also made important strides in wildlife medicine. Broken bones, injured wings, and severely damaged limbs have been returned to health using technologies adapted from pet and human medicine. A bald eagle shot by a poacher in Idaho in 2005 pro-

COMMON VETERINARY CHEMOTHERAPY DRUGS FOR CANCER

DRUG	TYPE OF CANCERS
chlorambucil (pill)	leukemia, lymphoma, multiple myeloma, ovarian cancer, polycythemia rubra vera
cisplatin (intravenous)	carcinomas (skin, ovarian, nasal, thyroid, mediastinal), osteosarcoma
cyclophosphamide (pill)	fast-growing cancers of bone marrow or lymphoma
doxorubicin (intravenous)	lymphoma, osteosarcoma, carcinomas, melanoma
L-asparaginase (intravenous)	lymphoma, mast cell tumors
lomustine (pill or intravenous)	skin lymphoma, tumors (mast cell, brain, kidney, lung), melanoma, histiocytic sarcoma
vincristine (intravenous)	lymphomas, mast cell tumors

vided veterinarians with an opportunity to find an innovative approach to trauma care. The poacher's shot had broken off the bird's upper beak and it had likely struggled to survive until someone found the starving bird scavenging for food in a landfill. The eagle was taken to a bird recovery center in Anchorage, Alaska, where the staff nursed her back to strength with hand-feeding, in the hope that a new beak would grow. But the beak did not grow, so engineer Nate Calvin of Boise, Idaho, who had heard the bird's story, set about to build a new beak. "As an engineer, as a human being first," he said to Associated Press reporter Nicholas Geranios, "I was interested in helping it out." Calvin modeled an artificial beak on a computer to attain the exact same dimensions as the natural beak. The resulting yellow nylon composite beak looked perfect but still had to be fixed onto Beauty (the eagle's new name). Calvin and a few volunteers glued a gold-titanium pin to the existing beak and then used the pin to help slide the new beak into place. After a bit of trimming and filing, Beauty used her prosthetic beak to grasp food on her own. Engineers and veterinarians soon designed a stronger, more permanent beak. Beauty has returned to Idaho where she serves as an ambassador for the inspiring advances that have become part of wildlife care.

CONCLUSION

Environmental toxins threaten wildlife in ways similar to the health threats they cause in humans. Wildlife confront additional threats because they have no choice but to drink tainted waters and breathe polluted air; they cannot seek out a doctor when they feel sick. Veterinarians and conservation biologists must therefore take the lead in finding wildlife that has been harmed by environmental hazards. Wildlife can then present these scientists with a variety of unique illnesses or injuries never seen before in medicine. In order to give wildlife adequate medical care, veterinary environmental medicine has taken cues from human medicine to diagnose, treat, and monitor disease. Veterinarians now use the most advanced medical techniques in caring for animals in zoos and rehabilitation centers. In past years a sick zoo animal would face euthanasia. Today wildlife care uses advanced medical instruments, drugs, and prostheses to give wild animals much longer life expectancy. These advances have come from veterinary medicine's willingness to try innovations that would never have been considered as little as a decade ago. Perhaps the dire situation of endangered species has led to the discovery of better medical care for animals in captivity.

Medical care for wildlife represents a final chance at saving some species. The main threat to wild animals comes from either towns or farmlands encroaching into wilderness. People bring pollution, wastes, roads, and noise with them. Some wildlife adapt to the turmoil but most species become vulnerable to new diseases and other threats. Any hope to conserve endangered species depends on a thorough understanding of these human-wildlife connections. Human activities probably cause hundreds of effects on wildlife that people do not realize. In subtle and in dramatic ways, biodiversity declines.

Veterinary environmental medicine therefore plays a critical role in protecting the diversity of life on Earth. Scientists in this field help animals by learning more about the toxins most dangerous to wildlife, the drugs and surgical methods that work best on exotic species, and improved methods in rehabilitation. Wildlife medical care has grown into one of the fastest advancing areas of environmental medicine.

FUTURE NEEDS

Illnesses in humans and animals caused by human-caused toxins have made environmental medicine increasingly important. This field of medicine examines the ailments caused by toxic chemicals and biological agents that are found in air, soil, or water. Environmental medicine also has been involved in testing the health effects of toxins in laboratories, monitoring organisms for their body content of chemicals, and developing better methods for treating humans or animals that have environmentally caused diseases.

Pollution is almost everywhere and it affects large populations. Scientists must use statistics to answer questions about entire populations by taking a sampling of the population. Biostatistics serve in epidemiology to find the source of a disease and to assess the spread of a disease through the population.

Environmental toxins belong to various categories based on chemical structure. Pesticides, organic compounds and solvents, heavy metals, and fine particles in the air are the primary health concerns in environmental medicine, though new threats have been gaining attention. The new environmental toxins that need more study are endocrine disrupters and chemicals used for making plastics. All of these substances create problems in outdoor air, indoor air, and in food and water.

When people who have existing health conditions are exposed to a toxin they have an added risk of becoming sick and developing more serious disease than the general population. Environmental medicine has the responsibility to pay particular attention to the needs of these at-risk individuals. Doctors must make use of all the new information on toxins in the environment as well as advanced medical technologies. People have limited options for preventing exposure to toxins, so medical doctors must also make their voices heard in issues of the environment that affect health: global warming, waste disposal, nuclear accidents, and hazardous waste spills. Epidemiologists can play an especially important role in government policymaking on pollution and climate change.

Environmental medicine therefore covers topics outside a doctor's office. This field reaches into larger concerns in environmental conservation. As long as chemicals and other substances threaten the well-being of people and animals in the environment, this medical specialty will become increasingly important.

Appendix A

Medical Meaning of Complete Blood Count (CBC) Measurements	
Blood Component	**Meaning**
red blood cells	number of red blood cells per volume of blood; increase or decrease may indicate abnormal conditions
hematocrit	percentage of red blood cells per volume of blood; may indicate anemia or other blood disorders
platelets	number of platelets per volume of blood; indicates normal bone marrow production of platelets and blood clotting ability
mean corpuscular volume (MCV)	average size of red blood cells; larger or smaller than normal indicates various nutrient deficiencies
white blood cells (WBC)	number of white blood cells per volume of blood; increase or decrease indicates abnormal conditions
sodium, potassium, chloride, calcium	electrolytes indicate proper hydration, heart rate, muscle function, blood pH
blood urea nitrogen (BUN)	indicator of kidney function
hemoglobin	indicator of anemia

Blood Component	Meaning
glucose	indicator of diabetes
total protein	indicator for kidney disease, liver disease, nutritional deficiencies
enzymes	liver function
high-density, low-density lipoproteins	blood transport of fats, possible indicator of heart disease
glomerular filtration rate (GFR)	flow rate of filtered fluid through the kidney, indicator of kidney function or disease
anion gap	difference between cations (positively charged ions) and anions (negatively charged ions) to measure acidity of blood, indicator of metabolic disorders

Appendix B

Medical Devices Used for Diagnosing Disease	
Instrument	**Use**
electrocardiogram (ECG or EKG)	analysis of electrical activity of the heart
computed tomography scan (CT scan or CAT scan)	detection of cancers, tumors, kidney failure, vascular disease
magnetic resonance imaging (MRI)	detection of tumors, heart and vascular diseases, liver lesions, abnormalities in reproductive organs
ultrasound (sonography)	detection of swelling, pain, or infection, examination of internal organs
X-ray (radiography)	detection of bone injuries/fractures, bone disease, fluid buildup, cancers, foreign objects
inductively coupled plasma mass spectrometry	measures trace (ppb–ppm) and ultratrace (ppq–ppb) elements
graphite furnace atomic absorption spectrometry	measures metals
isotope dilution mass spectrometry	measures organic compounds
Note: ppb = parts per billion; ppm = parts per million; ppq = parts per quadrillion	

Appendix C

Toxic Chemicals—Major Categories of the CDC Toxic Chemicals List	
Category	**Examples**
metals	mercury, lead, platinum
tobacco smoke	cotinine
polycyclic aromatic hydrocarbons (PAHs)	anthracene compounds
dioxins and furans	polychlorinated dibenzo-p-dioxins, polychlorinated dibenzofurans
non-dioxin polychlorinated biphenyls (PCBs)	pentachlorobiphenyl
phthalates	mono-methyl phthalate
phytoestrogens	daidzein
organochlorine pesticides	DDT, hexachlorobenzene, chlordane
organophosphate pesticides	dimethylphosphate
pyrethroid pesticides	phenoxybenzoic acid
herbicides	atrazine mercapturate
other pesticides	dichlorophenol
carbamate insecticides	carbofuranphenol

Appendix D

Examples of Toxic Organic Compounds and Solvents		
Compound	**Purpose**	**Human Health Concern**[1]
benzene, toluene, and other aromatic hydrocarbons	manufacture of dyes, detergents, paints, lacquers, adhesives, pesticides	neurological damage, cardiovascular damage
benzidene	manufacture of dyes	skin allergy, possible liver, neurological, and immune damage
bisphenol A	plastics	neurological effects, endocrine disorders
carbon tetrachloride and other chlorinated hydrocarbons	refrigerants, propellants, pesticides	liver, kidney, brain, and nerve damage
dioxins	by-products of pulp and paper industry, chemical manufacturing, chlorine disinfection of water and wastewater	skin diseases, hormone interference, weakened immune system, birth defects
ethylene oxide	sterilizes medical instruments, raw material in manufacturing	respiratory irritation, neurological damage, reproductive disorders
formaldehyde	paper and plywood production, preservative in cosmetics and medicines	eye, nose, and throat irritation, skin irritation

Compound	Purpose	Human Health Concern[1]
furans	by-products of pulp and paper industry, herbicide production	skin diseases, hormone interference, weakened immune system
hexane and other aliphatic hydrocarbons	solvent in manufacture of textiles, furniture, shoes	large amounts cause neurological and muscle disorders
perfluorinated alkylated substances (PFAs) such as perfluoroalkyloxy fluorocarbon	textiles, leathers, detergents, waxes	not fully determined
perfluorooctanoic acid and other perfluorinated compounds (PFCs)	nonstick cookware	possible reproductive and neurological effects
phthalates	plastics	reproductive disorders
polybrominated diphenyl ethers (PBDEs)	flame retardants for plastics and foams	affects thyroid gland function, possible liver damage, neurological and immune disorders
polychlorinated biphenyls (PCBs) (1977)[2]	coolants in electrical devices	liver damage, skin rashes, anemia
trichloroethylene (TCE)	cleaning solvent, paint removers, spot removers, adhesives	respiratory difficulty, neurological disorders, damage to cardiovascular system, liver damage
volatile organic compounds (VOCs)	paints, varnishes, wax, cleaning products	neurological damage and eyes, nose, and throat irritation

Notes:
[1] all of these compounds may cause cancers in addition to the diseases listed
[2] date in parentheses indicates year the compound was banned in the United States

Appendix E

Therapies Available for Exposure to Industrial Chemicals	
Chemical	**Therapy**
acids (sulfuric, phosphoric, nitric, hydrochloric, chromic)	water flushing of burned skin and eyes
acrylamide	wash skin with soap and water
bases (sodium hydroxide, potassium hydroxide, calcium oxide)	water flushing of burned skin and eyes; gentle rinsing with weak acid, such as 5% acetic acid
carbon disulfide	water flushing of skin and eyes
dioxins	none available
formaldehyde	water flushing of contaminated skin for 15 minutes then exposure to fresh air
hydrofluoric acid	water flushing of skin and eyes; neutralization with calcium gluconate solution
nitrogen aromatic compounds (benzidines, toluidines, anilines, naphthylamides)	administering methylene blue
nitrosamines	none available

Chemical	Therapy
polychlorinated biphenyls (PCBs)	wash skin with soap and water
polycyclic aromatic hydrocarbons (PAHs) (anthracene, pyrenes, naphthalene, bitumens)	treat dermatitis with cortisone creams
styrene	water flushing of contaminated skin for 15 minutes; no other treatments available
vinyl chloride	chemotherapy of liver cancer; no other treatments available

Glossary

ACUTE TOXICITY poisoning from a hazardous substance by receiving a large, single dose.

AIR QUALITY INDEX system for measuring the amount of unhealthy particles and gases in a set volume of air over a 24 hour period.

ALLERGEN a particle foreign to the body that causes an immune response in the body or on the skin.

AMBASSADOR an animal kept in captivity for use as an educational aid; usually an animal with an injury that prevents its safe return to the wild.

ANTIOXIDANTS substances that protect cells by neutralizing dangerous molecules called free radicals.

AT-RISK GROUP *also* high-risk group; a portion of the general population containing people with existing health conditions that make them more vulnerable to disease.

BEACH RESCUE saving an injured or diseased marine mammal stranded on a beach.

BIOACCUMULATION process in which the body stores toxins faster than it can degrade and excrete the toxins.

BIOASSAY test method to determine the effect of a toxin on a living organism.

BIOFUELS liquid or gas fuels made from plant material.

BIOMAGNIFICATION process of a toxin becoming more concentrated in tissues from the bottom to the top of a food chain.

BIOME a terrestrial area defined by the things living there, especially vegetation.

BIOMETRY statistics used in measurement studies of living systems.

BIOMONITORING procedure for determining the concentration of toxic chemicals in a person's or animal's body.

BIOPESTICIDE a substance made in nature that kills insects or other pests.

BIOSTATISTICS statistics used for studying populations of biological systems.

BIOTA living things—plant, animal, and microbial.

BIOTRANSFORMATION any enzymatic process that converts a harmful compound to a less harmful form.

CARBON CYCLE natural movement of different chemical forms of carbon from the atmosphere to land, through plant or animal life, and then back to the atmosphere.

CARBON RING chemical structure in which the carbon backbone forms a ring rather than a straight chain.

CARRYING CAPACITY maximum amount of organisms that an area can support over a given time period.

CHELATING AGENT compounds that capture metal molecules in a claw-like structure.

CHRONIC SUBLETHAL POISONING continual long-term exposure to a toxin that damages the body.

CHRONIC TOXICITY poisoning from a hazardous substance via a long-term continuous exposure or many repeated exposures.

CLEARANCE elimination of a substance from the blood by the kidneys.

CRITERIA POLLUTANTS five different chemicals and small particles in air for which the EPA sets strict standards because these pollutants are found throughout the United States.

CYTOTOXICITY poisoning of cells or structures within cells.

DEMOGRAPHICS study of a population's characteristics, such as age, gender, and ethnicity.

DESCRIPTIVE STATISTICS a method of analyzing a population's data to describe the main characteristics of the population.

DETOXIFICATION any process of neutralizing the hazardous effects of a toxic chemical.

DIAGNOSIS identification of an unknown disease.

DINOFLAGELLATE type of single-celled marine algae characterized by sturdy cell structure and the presence of two flagella.

DIRECT EFFECT symptoms that occur as a result of ingesting, inhaling, or absorbing a toxin.

DISINFECTION BY-PRODUCTS toxic chemicals formed during water disinfection.

DOSE the amount of a toxin taken into the body.

EMERGING DISEASE disease never seen before in medicine or never seen before in a species.

ENDOCRINE DISRUPTERS chemicals that interfere with normal hormone function in the body.

ENVIRONMENTALLY TRIGGERED ILLNESS (ETI) disease or injury that occurs when an environmental toxin affects the normal function of any body system or a person's comfort.

ENZYME a protein that aids chemical reactions in living things.

EPIDEMIOLOGY study of health and illness distributed across a population.

EPIGENETICS the study of gene groups that carry instructions for broad functions.

EPIGENOME a set of genes that controls the overall function of all other genes in the body but is not used for making new proteins.

FALLOUT radioactive dust and particles that fall out of the atmosphere.

FOUNDATION SPECIES species that creates or enhances habitat that benefits other species.

FREE RADICAL highly reactive intracellular molecule that makes other molecules reactive.

GAMMA RADIATION waves that are part of the electromagnetic spectrum at a specific frequency (about 10^{17} hertz) and caused by the energetic motions of electrons.

GASTROENTERITIS inflammation of the stomach and intestines with pain, cramping, and diarrhea.

GENE EXPRESSION synthesis of a protein that performs a job based on information carried in a cell's genes.

GENE POLYMORPHISM multiple version of a specific gene in a species' population.

GENETIC PREDISPOSITION above average likelihood of getting disease because of an individual's genetic makeup.

GENOME complete set of genetic information in an organism.

GREENHOUSE GAS atmospheric gas that participates in holding the Sun's energy in the troposphere in the form of heat.

HALONS hydrocarbons in which one or more hydrogen atoms is replaced by a halogen compound, such as bromine, chlorine, or fluorine.

HAY FEVER seasonal allergic reaction to particles such as pollen and characterized by sneezing, runny nose, and itchy, watery eyes.

HAZMAT an abbreviation for *hazardous materials,* substances that are toxic, flammable or ignitable, corrosive, highly reactive, explosive, or infectious.

HOMEOSTASIS process of the body to maintain a steady state.

IMMUNOCOMPROMISED condition of having a weakened immune system.

INDICATOR SPECIES species that give early warning of ecosystem destruction, usually by declining population.

INDIRECT EFFECT damages to health or symptoms in a person who has not come in contact with a toxin.

INFECTIOUS DISEASE illness caused by a microbe and capable of spreading among individuals.

INFERENTIAL STATISTICS a method of analyzing a population's data to allow scientists to draw conclusions about the population.

INVASIVE CANCER any cancer that spreads in the body; examples are cancers of the bladder, breast, colon and rectum, lung and bronchus, skin (melanoma), prostate, uterine cervix, uterine corpus, and leukemia and non-Hodgkin's lymphoma.

INVASIVE SPECIES a species foreign to a habitat, or nonnative.

IONIZING RADIATION alpha or beta particle emission or gamma rays released by radioactive chemicals.

ISOTOPES different forms of a chemical element that have the same number of protons but a dissimilar atomic mass because they have a different number of neutrons.

JAUNDICE yellowing of body tissues and skin due to buildup of liver secretions, mostly bile.

KEYSTONE SPECIES species upon which other species in an ecosystem depend.

LATENCY PERIOD time between toxin exposure and the onset of symptoms or disease.

LEVELS OF SIGNIFICANCE statistical values used to reject a null hypothesis, which is an assumption that states there is no difference between observed data and expected data.

MEDIAN LETHAL DOSE (LD_{50}) dose of toxin that kills 50 percent of test animals.

MESSENGER RIBONUCLEIC ACID (mRNA) type of RNA that controls the assembly of specific amino acids into proteins.

MINAMATA DISEASE severe mercury poisoning first seen in Minamata, Japan.

MIXED FUNCTION OXIDASES (MFO) enzymes that use oxygen as part of a reaction that converts two different compounds into another pair of compounds.

MULTIPLE CHEMICAL SENSITIVITY (MCS) complex set of symptoms caused by exposure to a mixture of metals, organic solvents, and pesticides.

NONPOINT SOURCES dispersed or large areas that release pollution into the environment.

NORMAL DISTRIBUTION symmetric dispersion of data around an average value; the bell curve.

NUTRIENT CYCLE *also* biogeochemical cycle; the reuse of elements and compounds by living and nonliving things on Earth.

OXIDATION chemical reaction in which a molecule loses electrons to another molecule.

OZONE SMOG irritating and unhealthy smog containing mostly ozone in the lower atmosphere.

OZONESONDE a balloon-borne instrument that measures ozone gas in the atmosphere either by detecting light wavelengths or chemical reactions.

PERSISTENCE period of time in which a contaminant stays in the environment or in the body.

PERSISTENT CHEMICALS chemicals that natural activities in the environment cannot break down or take a very long time to break down.

PHOTOCHEMICALS compounds produced when certain gases react in the atmosphere with sunlight.

PLASTICIZER substance in plastic that makes it flexible and durable.

POINT SOURCES specific, identifiable sites from which pollution originates and enters the environment.

P-VALUE the probability of observing a given result.

RADIOACTIVITY condition in which atoms become less energetically stable and as a consequence emit mass (alpha particles) or energy (gamma radiation) or both.

REGULATED CHEMICALS chemicals put into the environment by humans and known to cause harm to people or animals and which are limited in the environment by government agencies.

RIPARIAN pertaining to the specialized environment near and in rivers or streams.

RISK a way of describing the probability of becoming injured or sick.

SELECTIVE RESORPTION process by which kidneys excrete wastes and harmful chemicals by filtering the blood and retain useful compounds that the body can reuse.

SENTINEL SPECIES animal species that provide information about potential threats to human health.

SIGNIFICANCE a measure of the trustworthiness of sample data in describing a population from which the sample was taken.

SOLVENT any liquid used for dissolving a chemical; organic chemical solvents create air pollution and health hazards.

SPECTROMETER instrument that measures compounds or molecules by detecting characteristic light emission from the compound or molecule.

STANDARD upper allowable limit of a pollutant in air or water, enforced by U.S. law.

SURVIVORSHIP CURVE graphical depiction of the number of individuals expected to survive or die owing to a single factor.

SYNERGISM condition in which two chemicals or two processes together are more toxic than either one is by itself.

TEMPERATURE INVERSION weather condition in which a warm air layer rests atop a cool air layer nearer the ground, and the denser cool air traps pollutants in the low atmosphere.

THERMAL SHOCK condition in which an organism is critically or fatally hurt by sudden exposure to much warmer or much cooler temperatures than normal.

THERMOREGULATION natural process in the body to control body temperature.

TISSUE CULTURE technology in which living animal or plant cells are grown in a laboratory.

TOXICITY degree of being poisonous to living things.

TOXICOGENOMICS study of how a living thing's genes respond to environmental stress or toxins.

TOXICOKINETICS study of the movement of hazardous substances throughout the body.

TOXICOLOGY study of the detection and chemistry of toxic substances.

TOXICS RELEASE INVENTORY (TRI) U.S. Environmental Protection Agency (EPA) list of industrial chemicals that enter the environment, which identifies the chemical and the amount released.

TOXIN a poisonous substance of plant, animal, or microbial origin.

TRACE AMOUNTS concentration of any substance in the environment or in tissue at extremely low levels, usually measured at parts per million or lower.

TRACE-BACK following an illness or a pollutant back to its source.

TRACE GASES naturally occurring gases in the atmosphere at levels less than 1 percent of total gases.

TRANSBOUNDARY POLLUTION air or water pollution that originates in one place but pollutes another distant place, usually in another town, state, or country.

TRANSMITTANCE quality of a substance to let light pass through it; water has high transmittance.

ULTRAVIOLET RADIATION part of the light spectrum having energy higher than visible light and able to cause biological changes.

VECTOR insect that carries disease-causing microbes from one organism to another.

WATER RESCUE saving an injured or diseased marine mammal in deep or coastal waters.

WATER STRESS conditions in which a region's water requirements exceed water availability.

ZOOLOGICAL MEDICINE veterinary medicine pertaining to zoo animals or other exotic animals in sanctuaries or refuges.

ZOONOTIC DISEASE a disease that humans contract from animals.

Further Resources

Print and Internet

Adam, David. "World Carbon Dioxide Levels Highest for 650,000 Years, Says U.S. Report." *Guardian* (London), 13 May 2008. Available online. URL: www.guardian.co.uk/environment/2008/may/13/carbonemissions.climatechange. Accessed March 8, 2009. An update on the increased accumulation of atmospheric carbon dioxide.

Altman, Lawrence K. "In Philadelphia 30 Years Ago, an Eruption of Illness and Fear." *New York Times,* 1 August 2006. Available online. URL: www.nytimes.com/2006/08/01/health/01docs.html?_r=1&scp=1&sq=In+Philadelphia+30+Years+Ago%2C+an+Eruption+of+Illness+and+Fear&st=nyt. Accessed March 8, 2009. A recounting of the 1976 Legionnaires' disease outbreak as an example of indoor air pollution.

Associated Press. "Agency: Pollution Cutting Life Expectancy in Europe." *USA Today,* 10 October 2007. Available online. URL: www.usatoday.com/news/world/environment/2007-10-10-pollution-study_N.htm. Accessed March 8, 2009. A short article with an update on air and water pollution effects on European citizens' health.

———. "Intersex Fish Study Finds Endocrine Disruptors in Potomac Basin." 18 January 2007. Available online. URL: www.flmnh.ufl.edu/fish/InNews/intersex2007.html. Accessed March 8, 2009. A news release on newly discovered effects of endocrine disruptors on aquatic species.

Austen, Ian. "Canada Likely to Label Plastic Ingredient 'Toxic'." *New York Times,* 16 April 2008. Available online. URL: www.nytimes.com/2008/04/16/business/worldbusiness/16plastic.html?scp=1&sq=Canada%20Likely%20to%20Label%20Plastic%20Ingredient%20%91Toxic%92&st=cse. Accessed March 8, 2009. A recent article on Canada's approach to protecting against the plastic ingredient bisphenol A.

Balch, Phyllis A. *Prescription for Nutritional Healing.* 4th ed. New York: Avery, 2006. A book on holistic approaches to medicine containing opinions on environment contaminants.

Barlow, Jim. "Byproduct of Water-Disinfection Process Found to Be Highly Toxic." University of Illinois news release, 14 September 2004. Available online. URL: www.news.uiuc.edu/news/04/0914water.html. Accessed March 8, 2009. A news release on the toxic effects of water disinfection by-products.

Barrett, Stephen. "Exposing Multiple Chemical Sensitivity: Why This Diagnosis Is Spurious—And Why It Persists." *Nutrition Forum* 14, no. 2 (March/April 1997). Available online. URL: findarticles.com/p/articles/mi_m0GCU/is_n2_v14/ai_19513423. Accessed March 8, 2009. An article covering early debates on multiple chemical sensitivity to pollutants.

Bennett, Burton, Michael Repacholi, and Zhanat Carr, eds. *Health Effects of the Chernobyl Accident and Special Health Care Programmes.* Geneva, Switzerland: World Health Organization, 2006. Available online. URL: www.who.int/ionizing_radiation/chernobyl/WHO%20Report%20on%20Chernobyl%20Health %20Effects%20July%2006.pdf. Accessed March 8, 2009. A detailed report on the Chernobyl nuclear accident that serves as a resource for epidemiological and biological methods in environmental health.

Bio-Medicine. "Environmental Toxins May Cause Body's Defenses to Worsen Lung Disease," 6 September 2006. Available online. URL: www.bio-medicine.org/biology-news/Environmental-toxins-may-cause-bodys-defenses-to-worsen-lung-disease-4230-1. Accessed March 8, 2009. Summary of a university study on the effects of environmental chemical exposure on the body's immune system.

Blakemore, Bill. "People Sneezing and Wheezing—Climate Change at Work?" ABC News, 3 May 2007. Available online. URL: abcnews.go.com/Health/Allergies/story?id=3141554&page=1. Accessed March 8, 2009. A news story that covers the potential effect of global warming on higher pollen levels, which then increases respiratory disease.

Blumberg, Jeffrey, and Suzanne Le Quesne. "How Antioxidants Work." *Egyptian Doctor's Guide,* 2004. Available online. URL: www.drguide.mohp.gov.eg/newsite/News/HotTopics/Topic21.asp. Accessed March 8, 2009. A news article that describes the effects of free radicals in the body and how antioxidants work.

Bogden, John D., Mark A. Quinones, and Ahmed El Nakah. "Pesticide Exposure Among Migrant Workers in Southern New Jersey." *Bulletin of Environmental Contamination and Toxicology* 13, no. 5 (1975): 513–517. Available online. URL: www.springerlink.com/content/vp1125p08656gn07. Accessed March 8, 2009. A technical journal review article on pesticide exposure and effects in farmworkers.

British Broadcasting Company. "Cancer Alley, Louisiana, USA." BBC H2G2, 4 July 2002. Available online. URL: www.bbc.co.uk/dna/h2g2/A760420.

Accessed March 8, 2009. A short article describing the increased cancer rates along an industrial section of the Mississippi River.

———. "Ozone Hole Stable, Says Scientists." *BBC News,* 23 August 2006. Available online. URL: news.bbc.co.uk/2/hi/science/nature/5276994.stm. Accessed March 8, 2009. An update on the finding that the atmosphere's ozone depletion may have started to slow.

Carson, Rachel. *Silent Spring.* Boston: Houghton Mifflin, 1962. The seminal publication on pesticides in the environment and their effect on human and animal health.

Centers for Disease Control and Prevention. *Third National Report on Human Exposure to Environmental Chemicals.* Atlanta: CDC, 2005. Available online. URL: www.cdc.gov/exposurereport/report.htm. Accessed March 8, 2009. Biomonitoring results on the blood levels of environmental chemicals in U.S. adults and children.

Chameides, Bill. "The Link Between Environmental Toxins and Disease." *The Green Grok* blog, 11 June 2008. Available online. URL: www.nicholas.duke.edu/nicholas/insider/thegreengrok/epigenetics. Accessed March 8, 2009. Discussion on the potential role of environmental chemicals on gene function and disease.

Clark, Gregory. *A Farewell to Alms.* Princeton, N.J.: Princeton University Press, 2007. A book presenting the philosophy of economics, including a description of the Malthusian trap.

Cohn, Jeffery P. "The Comeback Condors." *Smithsonian Zoogoer,* March/April 2005. Available online. URL: nationalzoo.si.edu/Publications/ZooGoer/2005/2/condors.cfm. Accessed March 8, 2009. A well-written account of a project to save and rehabilitate endangered California condors.

Cone, Marla. "Chemical Exposure May Hurt Babies Later in Life." *San Francisco Chronicle,* 25 May 2007, A4. An article that covers an international scientist group's opinion on how environmental chemicals are affecting normal development in children.

———. "Waiting for the DDT Tide to Turn." *Los Angeles Times,* January 28, 2007. An article on the high levels of DDT still found in Pacific waters used for fishing.

Daily Telegraph (London). "Industrial Revolution: Survival of the Richest, Not the Fittest," 9 April 2007. Available online. URL: www.telegraph.co.uk/earth/main.jhtml?xml=/earth/2007/09/04/scposh104.xml&page=1. Accessed March 8, 2009. A British economist weighs the theories on what caused the Industrial Revolution and the consequences of industrialization.

Diamond, Jared. *Guns, Germs, and Steel: The Fates of Human Societies.* New York: W.W. Norton, 1997. An excellent, unique discussion on human relationships and science and how they shaped history.

Duncan, David Ewing. "The Pollution Within." *National Geographic,* October 2006. A detailed examination by the author on environmental contaminants being found in people's blood, including his own.

European Commission. *The Sixth Environment Action Programme of the European Community 2002–2012.* Available online. URL: ec.europa.eu/environment/newprg/intro.htm. Accessed March 8, 2009. An overview prepared by a committee of the European Union on biodiversity, climate change, natural resources, waste, and environmental health.

European Environment Agency. *Europe's Environment: The Fourth Assessment.* Copenhagen, Denmark: European Environment Agency, 2007. Available online. URL: www.eea.europa.eu/themes/regions/pan-european/fourth-assessment. Accessed March 8, 2009. A periodic report by the EEA to provide the latest scientific updates and study results on environmental health and other issues in environmental science.

Fahrenthold, David A. "Male Bass across Region Found to Be Bearing Eggs." *Washington Post,* 6 September 2006. Available online. URL: www.washingtonpost.com/wp-dyn/content/article/2006/09/05/AR2006090501384_pf.html. Accessed March 8, 2009. An article describing a local study on the effects of hormone disruptors on aquatic life.

Fountain, Henry. "Glaciers in Antarctica May Be Releasing DDT through Meltwater." *New York Times,* 27 May 2008. Available online. URL: www.nytimes.com/2008/05/27/science/earth/27obddt.html?scp=1&sq=Glaciers+in+Antarctica+May+Be+Releasing+DDT+Through+Meltwater&st=nyt. Accessed March 8, 2009. A study finds high levels of pesticides in Antarctica and in penguins' bodies.

Geranios, Nicholas K. "Eagle Wounded by Poacher Gets New Beak, New Look." *USA Today,* 6 June 2008. Available online. URL: my.att.net/s/editorial.dll?eeid=5916114&eetype=article&render=y&ck. A description of an unusual wildlife rehabilitation: a prosthetic beak for an injured eagle.

Greenpeace International. "And Justice Will Be Done?" *Greenpeace News,* 1 August 2003. Available online. URL: www.greenpeace.org/international/news/justice-for-warren-anderson. Accessed March 8, 2009. An account of the consequences of the Bhopal environmental disaster.

Guyton, Arthur C., and John E. Hall. *Textbook of Medical Physiology.* 11th ed. Philadelphia: W.B. Saunders, 2006. A detailed text on human physiology.

Hamilton, Doug. "Fooling with Nature: Interview with Jay Vroom." *PBS Frontline,* WGBH, March 1998. Available online. URL: www.pbs.org/wgbh/pages/frontline/shows/nature/interviews/vroom.html. Accessed March 8, 2009. The president of the American Crop Protection Association gives his views on the safety of pesticides.

Hotz, Robert Lee. "Asian Air Pollution Affects Our Weather." *Los Angeles Times,* 6 March 2007. Available online. URL: http://articles.latimes.com/2007/mar/06/science/sci-asiapollute6. Accessed March 8, 2009. An article describing how air pollution spreads across the globe.

Jehl, Douglas. "U.S. Going Empty-Handed to Meeting on Global Warming." *New York Times,* March 29, 2001. Available online. URL: http://query.nytimes.com/gst/fullpage.html?res=9504EEDA173FF93AA15750C0A9679C8B63&scp=1&sq=U.S.+Going+Empty-Handed+to+Meeting+on+Global+warming&st=nyt. Accessed March 8, 2009. This article provides a historical background on the United States' reaction to the Kyoto Protocol.

Kay, Jane. "Around the House: Indoor Air Pollution." *San Francisco Chronicle,* 19 May 2004. Available online. URL: www.sfgate.com/cgi-bin/article.cgi?file=/chronicle/archive/2004/05/19/HOGDC6LU141.DTL. Accessed March 8, 2009. An article on the many sources of dangerous indoor chemicals.

———. "State Senate Bill Would Ban Suspect Plastic." *San Francisco Chronicle,* 16 May 2008. Available online. URL: www.sfgate.com/cgi-bin/article.cgi?f=/c/a/2008/05/15/BAM610NCFR.DTL. Accessed March 8, 2009. An article on California's plan to ban the suspected endocrine disruptor bisphenol A.

———. "Toxic Toys." *San Francisco Chronicle,* 19 November 2006. An article on the unknown health effects of chemicals in plastics.

Kistner, William. "Environmental Chemicals and Human Health—New Links." *PBS Frontline Special Reports,* WGBH, 2005. Available online. URL: www.pbs.org/wgbh/pages/frontline/shows/nature/disrupt/envchem.html. Accessed March 8, 2009. The effects of exposure to environmental chemicals on children's IQ, behavior, and other health consequences.

Kolata, Gina. "Environment and Cancer: The Links Are Elusive." *New York Times,* 13 December 2005. Available online. URL: www.nytimes.com/2005/12/13/health/13canc.html?_r=1&scp=1&sq=Environment+and+Cancer%3A+The+Links+are+Elusive&st=nyt. Accessed March 8, 2009. An article that explains the medical profession's challenges in finding cause-and-effect between environmental chemicals and disease.

Kunzig, Robert. "Mopping Up the CO2 Deluge." *Time,* 3 July 2008. Available online. URL: www.time.com/time/magazine/article/0,9171,1820172,00.html. An interesting inspection into technologies for removing excess carbon dioxide from the atmosphere, including alterations to ocean chemistry. Accessed March 8, 2009.

Ladou, Joseph. *Current Occupational and Environmental Medicine.* 4th ed. New York: McGraw-Hill, 2007. A detailed resource on the physiological effects of environmental chemicals on the body.

Lawrence, Neal. "Do High-Voltage Power Lines Cause Cancer." *Midwest Today,* April/May 1996. Available online. URL: www.midtod.com/9603/voltage.phtml. Accessed March 8, 2009. An article from the early days of public concern over the health effects of living near power lines.

Lochhead, Carolyn. "Hundreds of Bird Species Decline as Habitat Is Lost." *San Francisco Chronicle,* 11 July 2008. An article updating the effects of habitat loss on bird numbers along a major migration corridor.

Lukas, David. "Blue Wilderness: Diving into Our Ocean Sanctuaries." *Bay Nature,* October/December 2007. Available online. URL: www.tmmc.org/pdfs/baynaturecolordec07.pdf. Accessed March 8, 2009. A well-written article on the environmental health of species that live along the U.S. Pacific coast.

Magnuson, Ed, J. Madeleine Nash, and Peter Stoler. "The Poisoning of America." *Time,* 22 September 1980. Available online. URL: www.time.com/time/magazine/article/0,9171,952748,00.html. Accessed March 8, 2009. An emotional account of the increasing amounts of contaminants entering the environment more than two decades ago.

Marsden, William. "We're in Chemical Overload." *Gazette* (Montreal), 20 June 2008. Available online. URL: www.canada.com/montrealgazette/story.html?id=e4c6d71f-2a6f-4952-98c7-24866f28aa67. Accessed March 8, 2009. An article on the troubling amounts of chemicals being found in people's bodies.

Martin, Eric P. "A Black Triangle Gradually Turns Green." Radio Prague, November 15, 2005. Available online. URL: www.radio.cz/en/article/72730. Accessed March 8, 2009. Transcript of a radio report on the Czech Republic's attempts to clean up one of Europe's most polluted regions.

McLamb, Eric. "The Industrial Revolution and Its Impact on Our Environment." *Ecology Global Network,* 19 May 2008. Available online. URL: www.ecology.com/features/industrial_revolution/index.html. Accessed March 8, 2009. A clear description of the relationships between industrialization, population growth, natural resources, and the future of the Earth's health.

McQuaid, John. "'Cancer Alley': Myth or Fact?" *New Orleans Times-Picayune,* May 23, 2000. Available online. URL: www.nola.com/speced/unwelcome/index.ssf?/speced/unwelcome/stories/0524b.html. Accessed March 8, 2009. An article that covers the difficulties and opinions that arise when attempting to draw a connection between pollution and cancer.

Medical News Today. "Childhood Leukaemia Risk Doubles Within 100 Metres of High Voltage Power Lines," 15 September 2004. Available online. URL: www.medicalnewstoday.com/articles/13440.php. Accessed March 8, 2009. A British 2004 article that discusses growing evidence on the health effects of high-voltage power lines.

———. "Environmental Toxin Causes Heritable Adult-Onset Diseases," 26 September 2006. Available online. URL: www.medicalnewstoday.com/articles/52628.php. Accessed March 8, 2009. An important article that describes the ways in which exposure to hazardous chemicals can cause health problems many years later in life.

NASA Goddard Space Flight Center. "New NASA/CSA Monitor Provides Global Air Pollution View from Space." Available online. URL: www.gsfc.nasa.gov/gsfc/earth/terra/co.htm. Accessed March 8, 2009. The technology of monitoring pollution by satellite imagery.

National Institute for Occupational Safety and Health. *NIOSH Pocket Guide to Chemical Hazards.* Available online. URL: www.cdc.gov/niosh/npg/default.html. Accessed March 8, 2009. A reference for information on environmental chemicals, including worker exposure protection and emergency procedures.

National Research Council of the National Academies. *Human Biomonitoring for Environmental Chemicals.* Washington, D.C.: National Academies Press, 2006. Available online. URL: http://books.nap.edu/openbook.php?record_id=11700&page=R1. Accessed December 6, 2008. A detailed and technical resource on biomonitoring methods, and background on this technique.

National Wildlife Federation. *Poisoning Wildlife: The Reality of Mercury Pollution.* September 2006. Available online. URL: www.nwf.org/wildlife/pdfs/Poisoningwildlife.pdf. Accessed March 8, 2009. A publication that summarizes the current knowledge (as of 2006) on mercury's health effects on aquatic, terrestrial, and bird species.

Nelson, Nancy. "The Majority of Cancers Are Linked to the Environment." National Cancer Institute *BenchMarks,* 17 June 2004. Available online. URL: www.cancer.gov/newscenter/benchmarks-vol4-issue3/page1. Accessed March 8, 2009. A clear review of the state of knowledge in 2004 on environmentally caused cancers.

Newman, Judith. "Allergy Misery: A Modern Epidemic." *National Geographic,* May 2006. Available online. URL: environment.nationalgeographic.com/environment/global-warming/allergies-misery.html. Accessed March 8, 2009. The author summarizes the current thinking on the rise of allergies.

Ogilvie, Felicity. "Penguin Droppings Help Identify Pesticide Hot Spots." Australian Broadcasting Corporation ABC News, 11 March 2008. Available online. URL: www.abc.net.au/news/stories/2008/03/11/2186867.htm. Accessed March 7, 2009. Evidence of the global spread of chemical pesticides to previously unpolluted places.

Pall, Martin L. "Live Chat with Martin L. Pall, PhD—July 6, 2007: Professor of Biochemistry Explains Mechanisms of Chronic Fatigue Syndrome and

Fibromyalgia and Suggested Protocol." July 9, 2007. Available online. URL: www.prohealthcom/library/showarticle.cfm?id/8145/&c+=CFIDS_FM. Accessed February 4, 2009. A discussion of the controversy of chronic fatigue syndrome and its suspected environmental causes.

PBS Newshour Extra. "World Health Organization Uses Controversial Insecticide to Combat Malaria," 18 September 2006. Available online. URL: www.pbs.org/newshour/extra/features/july-dec06/ddt_9-18.html. Accessed March 8, 2009. A report on the renewed use of DDT to fight global disease.

Prentice, Thomson, and Lina Tucker Reinders. *The World Health Report 2007: A Safer Future—Global Public Health Security in the 21st Century.* Edited by David Heymann. Geneva, Switzerland: World Health Organization, 2007. Available online. URL: www.who.int/whr/2007/en/index.html. Accessed March, 8, 2009. A comprehensive report on global health, including illnesses from poor water quality.

Prüss-Üstün, Annette, Robert Bos, Fiona Gore, and Jamoie Bartram. *Safer Water, Better Health.* Geneva, Switzerland: World Health Organization, 2008. Available online. URL: whqlibdoc.who.int/publications/2008/9789241596435_eng.pdf. Accessed March 8, 2009. WHO's regularly published update on world drinking water quality and sanitation.

Ramesh, Randeep. "Bhopal: Hundreds of New Victims are Born Each Year." *Guardian* (Manchester), 30 April 2008. Available online. URL: www.guardian.co.uk/world/2008/apr/30/india.pollution. Accessed June 25, 2009. An article describing the long-term health consequences of the 1984 chemical leak in Bhopal, India.

Roosens, Laurence, Nico Van Den Brink, Martin Riddle, Ronny Blust, Hugo Neels, and Adrian Covaci. "Penguin Colonies as Secondary Sources of Contamination with Persistent Organic Pollutants." *Journal of Environmental Monitoring* 9, no. 8 (August 2007): 822–825. Available online. URL: www.rsc.org/publishing/journals/EM/article.asp?doi=b708103k. Accessed March 8, 2009. A technical article on the pesticides being discovered in Antarctic penguins.

Rosenthal, Elisabeth. "Environmental Cost of Shipping Groceries around the World." *New York Times,* 26 April 2008. Available online. URL: www.nytimes.com/2008/04/26/business/worldbusiness/26food.html. Accessed March 8, 2009. The effects of global commerce in adding carbon dioxide to the atmosphere.

Rust, Susanne. "High Levels of Household Chemicals Found in Pets." *Milwaukee Journal Sentinel,* 17 April 2008. Available online. URL: www.redorbit.com/news/science/1348114/high_levels_of_household_chemicals_found_in_pets _study_results/index.html?source=r_science. Accessed March 8, 2009. An article that discusses the potential health threats to domestic animals from environmental chemicals.

Rutz, Dan. "Milwaukee Learned Its Water Lesson, But Many Other Cities Haven't." CNN.com, 2 September 1996. Available online. URL: www.cnn.com/HEALTH/9609/02/nfm/water.quality. Accessed March 8, 2009. An update on the quality of U.S. drinking water written shortly after a serious environmental contamination outbreak in Milwaukee.

Schmit, Julie. "U.S. Food Imports Outrun FDA Resources." *USA Today,* 18 March 2007. Available online. URL: www.usatoday.com/money/industries/food/2007-03-18-food-safety-usat_N.htm. Accessed March 8, 2009. A news article alerting the public to the potential biological and chemical contamination present in foods shipped around the world.

Schwartz, Noaki. "Endangered California Condors Turning Up with Lead Poisoning." *San Diego Union-Tribune,* 3 June 2008. Available online. URL: www.signonsandiego.com/news/state/20080603-1741-wst-poisonedcondors.html. Accessed March 8, 2009. Evidence of environmental contamination threatening the lives of a critically endangered species.

ScienceDaily. "Even Very Low Levels of Environmental Toxins Can Damage Health," 19 October 2005. Available online. URL: www.sciencedaily.com/releases/2005/10/051019233353.htm. Accessed March 8, 2009. Mounting evidence on the dangers of chemical exposure and human health.

———. "Male Bird at Smithsonian's National Zoo Has Reason to Celebrate Father's Day." June 15, 2008. Available online. URL: www.sciencedaily.com/releases/2008/06/080606160708.htm. Accessed March 8, 2009. A short article on zoo rehabilitation programs for endangered species, including the ostrichlike rhea.

Serrill, Michael S. "Anatomy of a Catastrophe." *Time,* 1 September 1986. Available online. URL: www.time.com/time/daily/chernobyl/860901.accident.html. Accessed March 8, 2009. An interesting recounting of a nuclear accident that polluted Chernobyl, Ukraine, written a few months after the event.

Shin, Annys. "Toxin Found in 'Natural,' 'Organic' Items." *Washington Post,* 15 March 2008. Available online. URL: www.washingtonpost.com/wp-dyn/content/article/2008/03/14/AR2008031403789.html. Accessed March 8, 2009. A report of the discovery by scientists of 1,4-dioxane in common consumer products.

Sink, Mindy. "Seeking Modern Refuge from Modern Life." *New York Times,* 19 October 2006. Available online. URL: www.nytimes.com/2006/10/19/us/19chemical.html?scp=1&sq=Seeking+Modern+Refuge+From+Modern+Life&st=nyt. Accessed March 8, 2009. An article on the controversial ailment called multiple chemical sensitivity syndrome.

U.S. Environmental Protection Agency. "DDT Ban Takes Effect." Press release, 31 December 1972. Available online. URL: www.epa.gov/history/topics/ddt/01.

htm. Accessed March 8, 2009. The EPA's original press release to announce the ban on DDT use in the United States.

——. *The Inside Story: A Guide to Indoor Air Quality.* Available online. URL: www.epa.gov/iaq/pubs/insidest.html. Accessed March 8, 2009. An informative resource on the sources of indoor air pollution, the main chemicals involved, and health issues from indoor pollution.

U.S. Geological Survey. "Pesticides in the Nation's Streams and Ground Water." News release, 3 March 2006. Available online. URL: www.usgs.gov/newsroom/article.asp?ID=1450. Accessed March 8, 2009. The summarization of a USGS report on waterway pollution.

Vachon, David. "Doctor John Snow Blames Water Pollution for Cholera Epidemic." *Old News* 16, no. 8 (May/June, 2005): 8–10. Available online. URL: www.ph.ucla.edu/epi/snow/fatherofepidemiology_part2.html#TWO. Accessed March 8, 2009. An interesting history of environmental pollution and epidemiology.

Vendantam, Shankar. "Kyoto Treaty Takes Effect Today." *Washington Post,* 16 February 2005. Available online. URL: www.washingtonpost.com/wp-dyn/articles/A27318-2005Feb15.html. Accessed March 8, 2009. A topical article written at the time of the Kyoto Treaty and the United States' decision in refusing to sign.

Vichit-Vadakan, Nuntavarn, Nitaya Vajanapoom, and Bart Ostro. "Public Health and Air Pollution in Asia (PAPA) Project: Estimating the Mortality Effects of Particulate Matter in Bangkok, Thailand." *Environmental Health Perspectives* 116, no. 9 (September 2008): 1,172–1,189. Available online. URL: www.ehponline.org/docs/2008/10849/abstract.html. Accessed March 8, 2009. A technical journal article on the amounts of air pollution in a large Asian city and its potential health effects on people.

Weise, Elizabeth, and Liz Szabo. "'Everywhere Chemicals' in Plastics Alarm Parents." *USA Today,* 3 October 2007. Available online. URL: www.usatoday.com/news/health/2007-10-30-plastics-cover_N.htm. Accessed March 8, 2009. An article on the occurrence of phthalates in beverage bottles for adults and infants.

Weiss, Rick. "Noise Pollution Takes Toll on Health and Happiness." *Washington Post,* 5 June 2007. Available online. URL: www.washingtonpost.com/wp-dyn/content/article/2007/06/04/AR2007060401430.html?sub=AR. Accessed March 8, 2009. An article summarizing recent scientific studies on the health effects of noise.

World Health Organization. *First World Health Assembly.* Official Records of the World Health Organization, No. 13, Plenary Meetings, Geneva, Switzerland, June 24–July 24, 1948. Available online. URL: http:whqlibdoc.who.int/hist/official_records/13e.pdf. Accessed March 8, 2009. The minutes of the first WHO assembly in 1948; excellent historical resource.

———. "Global Cancer Rates Could Increase by 50% to 15 Million by 2020." News release, 3 April 2003. Available online. URL: www.who.int/mediacentre/news/releases/2003/pr27/en. Accessed March 8, 2009. The WHO's periodic update on world cancer rates, with information on the general environmental and societal causes of cancer.

Zeman, Gary. "Health Risks Associated with Living Near High-Voltage Power Lines." *Health Physics Society,* 2 July 2008. Available online. URL: www.hps.org/hpspublications/articles/powerlines.html. Accessed March 8, 2009. An article giving a balanced discussion of the potential health risks of living near power lines.

Web Sites

American Academy of Environmental Medicine. URL: www.aaemonline.org. Accessed December 6, 2008. New concepts in blending Western medicines with holistic approaches to disease are explored, along with a good history of environmental medicine.

American Cancer Society. URL: www.cancer.org/docroot/home/index.asp. Accessed December 14, 2008. The primary resource for cancer information and causes, including updated cancer statistics.

Buck Institute for Age Research. URL: www.buckinstitute.org. Accessed December 6, 2008. An informative, technical site on the current research on aging and physiology.

Environmental Health Research Foundation Biomonitoring Info. URL: www.biomonitoringinfo.org/index.html. Accessed December 6, 2008. This site provides a clear description of biomonitoring and an excellent glossary.

National Cancer Institute. URL: www.cancer.gov. Accessed December 6, 2008. The NCI is the premier resource for information on various cancers and possible cancer causes, as well as health statistics.

National Institute of Environmental Health Sciences. URL: www.niehs.nih.gov/index.cfm. Accessed December 6, 2008. Current topics and breaking news in environmental health.

University of Minnesota College of Veterinary Medicine Raptor Center. URL: www.cvm.umn.edu/raptor/home.html. Accessed December 21, 2008. Background on raptor rehabilitation and ambassador birds.

U.S. Department of Health and Human Services. URL: www.hhs.gov. Accessed December 6, 2008. A government site providing an abundance of resources on health, with particular attention to biological agents.

World Health Organization. URL: www.who.int/en. Accessed December 6, 2008. The main resource for health issues worldwide.

Index

Note: Page numbers in *italic* refer to illustrations. The letter *t* indicates tables.

B

C

D

F

Y

Z

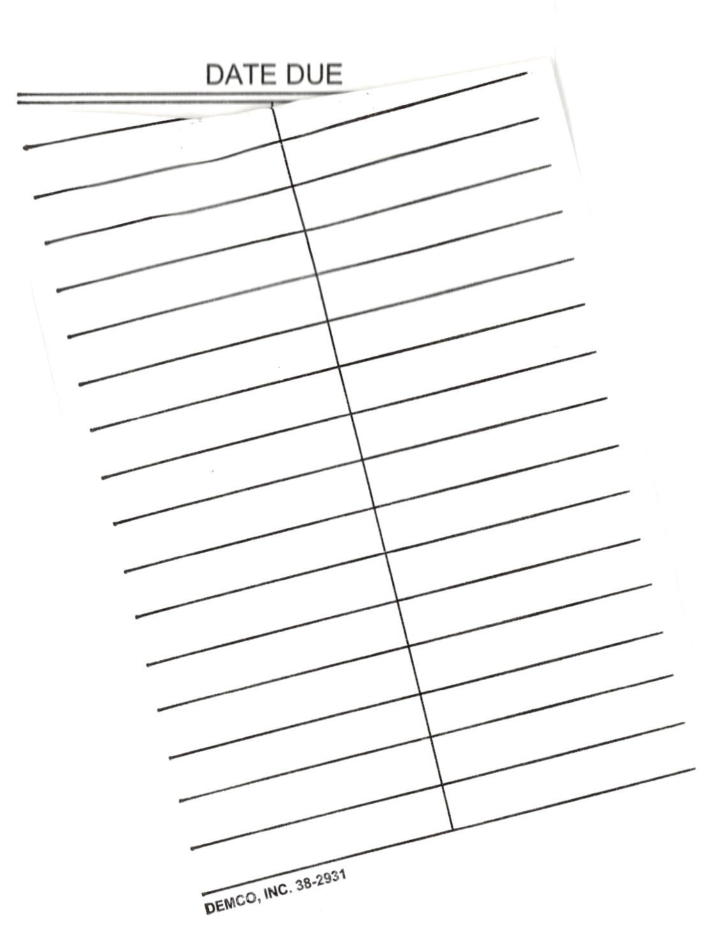